Quality Control and Assurance of the Deep Mixing Method

Quality Control and Assurance of the Deep Mixing Method

Masaki Kitazume
Tokyo Institute of Technology, Tokyo, Japan

CRC Press
Taylor & Francis Group
Boca Raton London New York

CRC Press is an imprint of the
Taylor & Francis Group, an **informa** business

Cover image: Author

First published 2022
by CRC Press/Balkema
Schipholweg 107C, 2316 XC Leiden, The Netherlands
e-mail: enquiries@taylorandfrancis.com
www.routledge.com – www.taylorandfrancis.com

CRC Press/Balkema is an imprint of the Taylor & Francis Group, an informa business

© 2022 Masaki Kitazume

The right of Masaki Kitazume to be identified as author of this work has been asserted in accordance with sections 77 and 78 of the Copyright, Designs and Patents Act 1988.

All rights reserved. No part of this book may be reprinted or reproduced or utilised in any form or by any electronic, mechanical, or other means, now known or hereafter invented, including photocopying and recording, or in any information storage or retrieval system, without permission in writing from the publishers.

Although all care is taken to ensure integrity and the quality of this publication and the information herein, no responsibility is assumed by the publishers nor the author for any damage to the property or persons as a result of operation or use of this publication and/or the information contained herein.

Library of Congress Cataloging-in-Publication Data
A catalog record has been requested for this book

ISBN: 978-1-032-12104-8 (hbk)
ISBN: 978-1-032-12108-6 (pbk)
ISBN: 978-1-003-22305-4 (ebk)

DOI: 10.1201/9781003223054

Typeset in Times New Roman
by codeMantra

Contents

Preface ix
Author xi
List of technical terms and symbols xiii
List of symbols xv

1 Overview of deep mixing method and scope of the book 1
  1.1 Definition of soft ground 1
  1.2 Outline of admixture stabilization 2
    1.2.1 Basic mechanism 2
    1.2.2 Type of admixture techniques 3
  1.3 Deep mixing method 4
    1.3.1 Outline of deep mixing method 4
    1.3.2 Classification of the deep mixing method 4
      1.3.2.1 On-land works 5
      1.3.2.2 Marine works 7
    1.3.3 Column/element installation patterns and applications 9
  1.4 Scope of book 11
  References 12

2 Quality control and assurance of deep mixing method 15
  2.1 Importance of quality control and quality assurance 15
  2.2 Work flow of deep mixing project and QC/QA 15
  2.3 Current practice of QC/QA 20
    2.3.1 Basic concept of laboratory, field and design standard strengths 20
    2.3.2 Process design 21
      2.3.2.1 Flow of mixing design and process design 22
      2.3.2.2 Mixing condition in laboratory and field 23
      2.3.2.3 Tips of laboratory mix test 25
    2.3.3 Selection of deep mixing equipment 26
    2.3.4 Field-trial test 27
    2.3.5 Quality control during production 28
      2.3.5.1 Construction procedure 28
      2.3.5.2 Overlap columns/elements 30

		2.3.5.3	Operational parameters	32
		2.3.5.4	Example of construction procedure	35
	2.3.6	Quality control throughout construction period		36
		2.3.6.1	Material management	36
		2.3.6.2	Modification of construction control values	37
		2.3.6.3	Damage of mixing tool	37
		2.3.6.4	Lateral displacement and ground heaving	37
	2.3.7	Report		40
	2.3.8	Quality verification		44
		2.3.8.1	Verification methods	44
		2.3.8.2	Position of core boring	44
		2.3.8.3	Frequency of core boring	45
		2.3.8.4	Quality verification of boring core sample	46
		2.3.8.5	Quality verification by laboratory test	47
		2.3.8.6	Evaluation of unconfined compressive strength	48
	2.3.9	Rectification of non-compliant column/element		49
References				*50*
3	**Technical issues on QC/QA of stabilized soil**			**53**
3.1	*Introduction*			*53*
3.2	*Field and laboratory strengths*			*53*
	3.2.1	Prediction of strength		53
	3.2.2	Strength ratio of field to laboratory strengths, q_{uf}/q_{ul}		54
	3.2.3	Strength deviation in field strength		55
3.3	*Laboratory mix test*			*57*
	3.3.1	Role and basic approach of laboratory mix test		57
	3.3.2	Selection of soil for laboratory test and water to binder ratio of binder slurry, w/c		58
	3.3.3	Effect of specimen size		59
		3.3.3.1	Strength	60
		3.3.3.2	Young's modulus	62
	3.3.4	Effect of molding technique		62
	3.3.5	Effect of overburden pressure during curing		65
	3.3.6	Effect of curing temperature		67
		3.3.6.1	Temperature in ground	67
		3.3.6.2	Effects of curing temperature and period	68
		3.3.6.3	Maturity	71
3.4	*Selection of deep mixing equipment*			*73*
	3.4.1	Factors influencing mixing degree		73
		3.4.1.1	Influence of number of mixing shafts	73
		3.4.1.2	Influence of type and shape of mixing blade	73
		3.4.1.3	Influence of diameter of mixing blade	76
		3.4.1.4	Influence of penetration speed of mixing tool	77

	3.4.2	Required blade rotation number		78
		3.4.2.1	Influence of blade rotation number in laboratory model tests	78
		3.4.2.2	Influence of blade rotation number in field test	79
		3.4.2.3	Influence of blade rotation number in field actual works	80
	3.4.3	Stabilization at shallow depth and influence of ground heaving		81
		3.4.3.1	Basic production procedure and effect of sand mat	81
		3.4.3.2	Influence of ground heaving	82
	3.4.4	Bottom treatment		85
	3.4.5	Overlap columns/elements		86
3.5	*Verification techniques in quality assurance*			*88*
	3.5.1	Core boring		88
		3.5.1.1	Procedure	88
		3.5.1.2	Frequency of boring core sampling and specimen	89
		3.5.1.3	Coring boring technique	89
		3.5.1.4	Size of boring core	90
		3.5.1.5	Macroscopic evaluation of strength of field-stabilized soil	91
	3.5.2	Applicability of wet grab sampling		94
		3.5.2.1	Type of wet grab sampling	94
		3.5.2.2	Comparison of sampling type	96
		3.5.2.3	Comparison of wet grab sample strength and boring core sample strength	97
		3.5.2.4	Applicability of wet grab sampling for QA	98
References				*99*

4 Problems and countermeasures associated with problematic soils — 105

4.1	*Problematic soil for stabilization*		*105*
4.2	*Countermeasures for problematic soils*		*105*
	4.2.1	Water injection	105
	4.2.2	Use of new type special cement	106
	4.2.3	Use of dispersant	107
	4.2.4	Injecting atomized cement slurry	110
	4.2.5	Summary	111
References			*111*

5 Water to binder ratio concept in QC — 113

5.1	*Introduction*		*113*
5.2	*Definition of W/C ratio*		*113*
	5.2.1	Definition of *W/C*	*113*
	5.2.2	Relationship between *W/C* ratio and stabilized soil strength	114
5.3	*Prediction of field strength by production log data*		*115*

	5.3.1	Production log data	115
	5.3.2	Analysis of production log data	117
	5.3.3	Countermeasure for water injection	120
5.4	Summary		122
References			123

Index   *125*

Preface

The deep mixing method, a deep in-situ admixture stabilization technique using lime, cement or lime-based and cement-based special binders, was developed in Japan and Sweden in the 1970s. Compared to the other ground improvement techniques, deep mixing has advantages such as the large strength increase within a month period, little adverse impact on environment and high applicability to any kind of soil if binder type and mount are properly selected. The application covers on-land and in-water constructions ranging from strengthening the foundation ground of buildings, embankment supports, earth retaining structures, retrofit and renovation of urban infrastructures, liquefaction hazards mitigation, man-made island constructions and seepage control. Until the end of the 1980s, the deep mixing method has been developed and practiced only in Japan and Nordic countries with a few exceptions. In the 1990s, the deep mixing method gained popularity also in the United States of America and central Europe.

Improved ground by the method is a composite system comprising stiff stabilized soil and unstabilized soft soil, which necessitates geotechnical engineers to fully understand the interaction of stabilized and unstabilized soil and the engineering characteristics of in-situ stabilized soil. Based on the knowledge, the geotechnical engineer determines the geometry (plan layout, verticality and depth) of stabilized soil elements, by assuming/establishing the engineering properties of stabilized soil, so that the improved ground may satisfy the performance criteria of the superstructure. The success of the project, however, cannot be achieved by the well-determined geotechnical design alone. The success is guaranteed only when the quality and geometric layout envisaged in the design is realized in the field with an acceptable level of accuracy.

The strength of the stabilized soil is influenced by many factors including original soil properties and stratification, type and amount of binder, curing conditions and mixing process. The accuracy of the geometric layout heavily depends upon the capability of mixing equipment, mixing process and contractor's skill. Therefore, the process design, production with careful quality control (QC) and quality assurance (QA) are key to the deep mixing project. Quality control and assurance starts with the soil characterization of the original soil and includes various activities prior to, during and after the production. QC/QA methods and procedures and acceptance criteria should be determined before the actual production and their meanings should be understood precisely by all the parties involved in a deep mixing project.

The design, construction and quality control and assurance were well established so that the method has been frequently applied in construction of infrastructures worldwide. Many design manuals provide fundamental design concept and procedures for the improved ground based on very simple assumptions and idealization of the improved ground. The recent development of computers and numerical analyses, on the other hand, enabled a very sophisticated and comprehensive design of the improved ground, in which the improved ground is modeled as an elast-plastic material such as Cam Clay model. The more constitutive modeling on the improved soil requires many soil parameters that in turn more comprehensive tests on the improved soil. However, the improved soil is not homogeneous and the surrounding soil is disturbed and deformed due the construction of the improved soil. The computer-based design cannot achieve high accurate design unless it incorporates these effects precisely. The comprehensive design sometimes requires complicated construction, such as honeycomb-shaped improved ground that requires exact overlapping with many improved columns.

I had many opportunities to be engaged in domestic and foreign deep mixing projects. I found some troubles and problems in the construction that could not satisfy the criteria and performance specified in the design. The reasons behind them are not always the insufficient knowledge and ability of constructors but sometimes the design that does not consider the capacity and ability of deep mixing equipment and procedure. The consultants should study the construction equipment, production process and quality control and assurance of the method in order to make appropriate design incorporating the capacity and ability of the equipment and also the ground conditions and to achieve the design criteria and performance of the improved ground specified in the design.

The current book is intended to provide the state of the art and practice of quality control and assurance on deep mixing rather than a user-friendly manual. Many books are published on the deep mixing method that covers various topics regarding the deep mixing briefly, such as the factors affecting the strength increase, the engineering characteristics of stabilized soil, a variety of applications and associated column installation patterns, current design procedures, execution systems and procedures, and QC/QA methods and procedures. The book covers the quality control and assurance only but in detail based on the experience and research efforts accumulated in the past 50 years in Japan in order to provide solutions and countermeasures to any troubles or problems encountered at field.

The author wish the book to be useful for practicing engineers to understand the current state of the art of the QC/QA of the deep mixing method and also useful for academia to find out the issues to be studied in the future.

June 2021

Author

Masaki Kitazume, graduated from Tokyo Institute of Technology in 1979, obtained his Master of Engineering in 1981. Then, he joined the Port and Harbour Research Institute, Ministry of Transport and has been the head of the Soil Stabilization Laboratory and has worked on the interaction of improved ground and soft ground. In 1994, he got a Doctor of Engineering from Tokyo Institute of Technology on the design of stability of Deep Mixing improved ground. In 2011, he was invited to become a professor of the Department of Civil and Environmental Engineering, Tokyo Institute of Technology.

He has published many papers, mainly on the geotechnical aspects of soil stabilization, ground improvement and centrifuge model testing. He also published three books from Balkema Publishers and Taylor & Francis, on *Deep Mixing Method*, *Sand Compaction Pile Method* and *Pneumatic Flow Mixing Method*.

He was awarded the Geotechnical Engineering Development award from the Japanese Society of Soil Mechanics and Foundation Engineering in 1992, the Minister of Transport Award in 2000, and Continuing International Contribution Awards, Japan Society of Civil Engineers in 2006, Geotechnical Engineering Research Achievements Award, the Japanese Society of Soil Mechanics and Foundation Engineering in 2018, and Telford Premium Prize, ICE Awards in 2019.

List of technical terms and symbols

Definition of technical terms

additive: chemical material to be added to stabilizing agent for improving characteristics of stabilized soil

beep mixing equipment: deep mixing equipment with various mixing tools including vertical shaft mixing tools and horizontal rotating circular cutters

binder: chemically reactive material (i.e., lime, cement, gypsum, blast furnace slag, flyash, or other hardening reagents)

binder content: ratio of weight of dry binder to dry weight of soil to be stabilized

binder factor: ratio of weight of dry binder to volume of soil to be stabilized

binder slurry: slurry-like mixture of binder and water

blade rotation number: total number of mixing blade rotations per meter of shaft movement

cement content: ratio of weight of dry cement to dry weight of soil to be stabilized

cement factor: ratio of weight of dry cement to volume of soil to be stabilized

cement slurry: slurry-like mixture of cement and water

deep mixing construction equipment: a sort of deep mixing system consists of mixing equipment, binder plant and control units, etc.

external stability: overall stability of the stabilized body

field strength: strength of stabilized soil produced in-situ

fixed type: a type of improvement in which stabilized soil column reaches a bearing layer

floating type: a type of improvement in which stabilized soil column ends in a soft soil layer

improved ground: a region with stabilized soil columns/elements and surrounding original soil

internal stability: stability on internal failure of improved ground

laboratory strength: strength of stabilized soil produced in a laboratory

stabilized soil: soil stabilized by mixing with binder

stabilized soil column: column of stabilized soil produced by a single shaft mixing tool

stabilized soil element: element of stabilized soil produced by a multipleshaft mixing tool water to binder ratio of binder slurry ratio of weight of dry binder to weight of water of binder slurry total water to binder ratio: ratio of weight of dry binder to total weights of water of binder slurry and soil water to cement ratio of cement slurry ratio of weight of dry cement to weight of water of cement slurry

total water to cement ratio: ratio of weight of dry cement to total weights of water of cement slurry and soil

List of symbols

aw:	binder content, cement content
COV:	coefficient of variation
E_{50}:	elastic modulus (kN/m^2)
E_{50}/q_u:	ratio of elastic modulus to unconfined compressive strength
Fc:	binder factor, cement factor (kg/m^3)
M:	maturity (°C—day)
N_d:	number of rotation of mixing tool during penetration (N/min)
N_u:	number of rotation of mixing tool during withdrawal (N/min)
q_u:	unconfined compressive strength (kN/m^2)
q_{uck}:	design standard strength (kN/m^2)
q_{uf}:	unconfined compressive strength of field stabilized soil (kN/m^2)
$\overline{q_{uf}}$:	average unconfined compressive strength of field stabilized soil (kN/m^2)
q_{ul}:	unconfined compressive strength of laboratory stabilized soil (kN/m^2)
$\overline{q_{ul}}$:	average unconfined compressive strength of laboratory stabilized soil (kN/m^2)
RQD:	rock quality designation index
T:	blade rotation number (N/m)
T_0:	reference temperature (–10°C)
t_c:	curing period (day)
T_c:	curing temperature (°C)
V_d:	penetration speed of mixing shaft (m/min)
V_u:	withdrawal speed of mixing shaft (m/min)
w/c:	water to binder ratio of binder slurry, water to cement ratio of cement slurry
W/C:	total water to binder ratio of stabilized soil, total water to cement ratio of stabilized soil
λ:	strength ratio of q_{uf}/q_{ul}
ΣM:	total number of mixing blades

Chapter 1

Overview of deep mixing method and scope of the book

1.1 DEFINITION OF SOFT GROUND

It becomes difficult to locate a new infrastructure on a stiff and hard ground in urban areas throughout the world. Renovation or retrofit of old infrastructures should often be carried out in the close proximity of the existing structures. Good-quality ground for constructions is becoming a precious resource to be left for the next generation. Due to these reasons and environmental restrictions on the public works, ground improvement is becoming an essential part of the infrastructure development projects both in the developed and developing countries. This situation is especially pronounced in Japan, where many construction projects must locate on soft alluvial clay grounds, artificial lands reclaimed with soft dredged soils, highly organic soils and so on. These ground conditions would pose serious problems of large ground settlement and/or instability of structures. Apart from clayey or highly organic soils, loose sand deposits under water table would cause a serious problem of liquefaction under seismic condition. When these problems are anticipated not to assure the performance and function of superstructure, the ground is called a 'soft ground' and needs to be improved and reinforced. Required performance and function of the ground are, however, different for different structures. It is not appropriate to define a 'soft ground' by its geotechnical characteristics alone but by incorporating the size, type, function and importance of superstructure and construction period. Only if the type of superstructure is specified, it is possible to define the 'soft ground.' Table 1.1 provides rough idea of the 'soft ground' for several types of structure in terms of water content, unconfined

Table 1.1 Definition of soft ground for several types of structure (The Japanese Society of Soil Mechanics and Foundation Engineering, 1986).

	Highway			Railway		Building	Fill dam
	Water content (%)	UCS, q_u (kN/m²)	SPT N-value	SPT N-value	Thickness (m)	Bearing capacity (kN/m2)	SPT N-value
Organic soil	>100	<50	<4	0	>2	<100	<20
Clayey soil	>50	<50	<4	2	>5	<100	-
Sandy soil	>30	≒ 0	<10	4	>10	-	-

DOI: 10.1201/9781003223054-1

compressive strength, SPT N-value, ground thickness and bearing capacity (The Japanese Society of Soil Mechanics and Foundation Engineering, 1986).

If superstructure to be constructed would be unstable under given conditions of external loads and of original ground, or if expected deformation during and/or after construction would exceed an allowable value from the viewpoint of expected function of superstructure, necessary countermeasures must be undertaken. The following four approaches can be applied: (a) changing type of superstructure and/or type of its foundation, (b) replacing soft soil by better quality soil, (c) improving properties of soft soil, and (d) introducing reinforcing material into soft soil. 'Ground improvement' covers (b), (c) and (d) above, and can be defined as any countermeasures given to soft soil in order to attain the successful performance of superstructure if otherwise unattainable. The ground improvement techniques can be classified, based on their working principles, into replacement, densification, consolidation/dewatering, grouting, admixture stabilization, thermal stabilization, reinforcement and miscellaneous. These techniques have been introduced to or originally developed in Japan during the past decades.

The deep mixing method, one of the admixture stabilization techniques, was developed in Japan and put into practice in the middle of the 1970s. Since then, the wet and dry methods have been applied to many improvement purposes and a lot of research studies and case histories have been accumulated (Kitazume and Terashi, 2013).

1.2 OUTLINE OF ADMIXTURE STABILIZATION

1.2.1 Basic mechanism

Admixture stabilization is a technique of mixing chemical binder with soil to improve the consistency, strength, deformation characteristics and permeability of soil. When, for example, cement absorbs the pore water in the soil, cement mineral, $3CaO.SiO_2$, for example, reacts with water in the following way to produce cement hydration products, Equation (1.1).

$$2(3CaO \cdot SiO_2) + 6H_2O = 3CaO \cdot 2SiO_2 \cdot 3H_2O + 3Ca(OH)_2 \qquad (1.1)$$

During the cement hydration, calcium hydroxide, $Ca(OH)_2$, is released. The cement hydration product has high strength, which increases as it ages, while calcium hydroxide contributes to the pozzolanic reaction. The improvement becomes possible by the ion exchange at the surface of clay minerals, bonding of soil particles and/or filling of void spaces by chemical reaction products. Although a variety of chemical binders have been developed and used for the admixture stabilization, the most frequently used binders nowadays are lime and cement due to their availability and cost. The mechanisms of the lime and cement stabilizations were studied in the 1960s by the highway engineers in relation to the improvement of base and sub-base materials for road construction (Ingles and Metcalf, 1972). The physical and engineering properties of lime and cement-stabilized soil have been studied extensively since then. The rather complicated mechanism of cement stabilization is simplified and schematically shown in Figure 1.1 for the chemical reactions between clay, pore water, cement and slag (Saitoh et al., 1985).

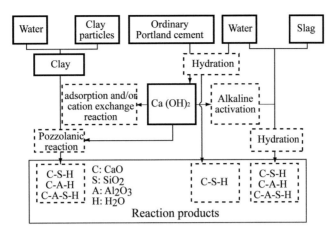

Figure 1.1 Chemical reactions between clay, cement, slag and water (Saitoh et al., 1985).

1.2.2 Type of admixture techniques

Many types of admixture stabilization techniques have been developed in Japan, which can be classified into the *in-situ* mixing and the *ex-situ* mixing, as shown in Table 1.2 (after Coastal Development Institute of Technology, 2008). The *in-situ* mixing techniques are developed to improve the physical and mechanical properties of

Table 1.2 Classification of admixture stabilization techniques after Coastal Development Institute of Technology (2008).

Place of mixing		Type of mixing	Method	Application
In-situ	Surface and shallow depth stabilization	Mechanical mixing	Surface treatment, shallow stabilization	Working platform on soft ground
	Mid-depth stabilization	Mechanical mixing	Mid-depth mixing	Stability, settlement reduction, excavation support, seepage shutoff, etc.
	Deep depth stabilization	Combination of mechanical and high-pressure injection	Deep mixing	
Ex-situ	Mixing during transportation	Mixing on belt conveyor	Pre-mixing	Improve liquefaction resistance of soil
		Mixing in pipeline	Pipe mixing	Reduce compressibility of high-water-content soil
	Mixing in batch plant	Mechanical mixing	Pre-mixing	Improve liquefaction resistance of soil
		Mechanical mixing	Lightweight geo-material	Density control of fill material
		Mechanical mixing and high-pressure dewatering	Dewatered stabilized soil	Alternative for sand and gravel

original soil for assuring the successful performance of superstructure on a ground, where original soil is stabilized with the chemical binder *in-situ* by means of mechanical mixing and/or high-pressure injection mixing. The *in-situ* mixing techniques can be subdivided into surface and shallow depth stabilization, mid-depth stabilization, and deep depth stabilization principally depending upon the depth and purpose of improvement. The *ex-situ* mixing techniques have been developed to enhance the beneficial use of dredged soils, inappropriate soils and construction surplus soils. These techniques are intended to provide additional characteristics to an original soil, such as better liquefaction resistance, smaller density, smaller volume compressibility or high strength. In the *ex-situ* mixing techniques, the soils are once excavated or collected, mixed with the chemical binder in a plant or in a transportation pipeline or at a reclamation site. The *ex-situ* mixing techniques can be further classified into the mixing during transportation and batch plant mixing, depending upon where soil and binder are mixed.

1.3 DEEP MIXING METHOD

1.3.1 Outline of deep mixing method

Deep mixing method is categorized into the *in-situ* mixing technique, where a large size equipment having rotating mixing blades is used for supplying chemical binder into a ground to produce a column shape stabilized soil in-situ. The strength of stabilized soil by the method is of the order of 500–5,000 kN/m^2 in terms of unconfined compressive strength. This method was originally developed in the 1970s and has frequently applied to on-land and marine works in Japan. The method has been described in detail (Coastal Development Institute of Technology, 2019; Kitazume and Terashi, 2013). The lime column method, the other type of the deep mixing method, was developed in Sweden in the same period (Broms, 1984, 1991), which exhibits the same technology in principle.

The extensive investigations have been carried out in Japan on the strength increase by lime and cement, the engineering characteristics of stabilized soil, the equipment and execution system, the geotechnical design and the quality control and quality assurance (Kitazume, 2011; Terashi and Kitazume, 2009, 2011, 2015)

1.3.2 Classification of the deep mixing method

The techniques for the deep mixing method can be divided into three groups: mechanical mixing, high-pressure injection mixing, and combination of mechanical and high-pressure injection mixing. The various methods in these groups are classified in Figures 1.2 and 1.3, in which the most common methods used in Japan are listed. In the mechanical mixing technique, binder is fed to soft ground and forcibly mixed with the in-situ soil by the mixing tool. The chemical binder is used either with a slurry form or with a dry form. The Cement Deep Mixing (CDM) method, the most common wet method in Japan, has been frequently applied for both marine and on-land works (Cement Deep Mixing Method Association, 1999). The Dry Jet Mixing (DJM) method

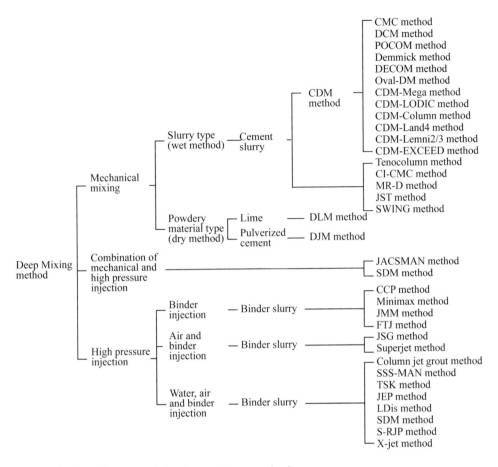

Figure 1.2 Classification of the deep mixing method.

is the most common dry method in Japan and is usually applied for land construction (Dry Jet Mixing Method Association, 2006; Public Works Research Center, 2004). In the high-pressure injection technique, on the other hand, the original ground is disturbed and softened with high-pressure jet of water and/or air, while at the same time, the chemical binder is injected into the ground and mixed with the in-situ soil. In the combination of mechanical and high-pressure injection mixing, the original ground is disturbed and softened by the mixing blades and high-pressure jet, JACSMAN (Jet And Churning System MANagement) (Miyoshi and Hirayama, 1996; JACSMAN Association, 2011).

1.3.2.1 On-land works

There are basically two types of deep mixing equipment for on-land works; large size equipment for civil engineering facilities (Cement Deep Mixing Method Association, 1999; Dry Jet Mixing Method Association, 2006) and small size equipment for

Figure 1.3 Classification of deep mixing method. (a) Mechanical mixing. (b) High pressure injection. (c) Combination of mechanical and high pressure injection (By the courtesy of Fudo Tetra Corporation).

residential houses (The Building Center of Japan and the Center for Better Living, 2018). In the former, deep mixing equipment usually has one or two mixing shafts (Figures 1.4 and 1.5). The mixing blade has a diameter of about 1–1.6m, which can make a column/element with a cross-sectional area ranging from 0.8 to 2.0 m^2. The maximum depth of stabilization reaches down to 40m depth.

A two mixing shafts deep mixing equipment usually has a bracing plate to keep the distance of the two mixing shafts. The plate is also expected to function to increase the mixing degree by preventing the entrained mixing phenomenon, a condition in which disturbed soft soil adheres to and rotates with the mixing blade without efficient mixing of the binder and soil.

Figure 1.4 Mechanical deep mixing equipment for on-land works.

For a single mixing shaft equipment, a free blade, an extra blade about 10 cm longer than the mixing blade, is usually installed close to the mixing blade to prevent the entrained mixing phenomenon. In the latter equipment, for foundation of residential house, deep mixing equipment usually has one mixing shaft (Figures 1.6a). The mixing blade is rather simple having a diameter of about 0.6 m (Figure 1.6b), which can make a stabilized soil column with a cross-sectional area of around $0.3\,\text{m}^2$. The maximum depth of the stabilized soil column is about 13 m depth. A free blade is usually installed close to the mixing blade to prevent the entrained mixing phenomenon.

1.3.2.2 Marine works

In marine works, a special barge installed with deep mixing equipment is used to improve soft soil *in-situ* (Figure 1.7a). The deep mixing equipment for marine work

Figure 1.5 Mixing blades of deep mixing equipment for on-land works.

Figure 1.6 Small size deep mixing equipment for on-land works.

Figure 1.7 Deep mixing equipment for marine works.

usually have more than two mixing shafts, as shown in Figure 1.7b. The deep mixing equipment currently available in Japan are capable of constructing large stabilized soil element whose cross-sectional area is ranging from 1.5 to 9.5 m^2 and the maximum depth of stabilization reaches down to −70 m from the water surface. The deep mixing equipment can penetrate the local stiff layer to reach the design depth. An equipment with a relatively large capacity can penetrate a layer whose SPT N-value and thickness are 8 and 4 m for clayey soil, and 15 and 4 m for sandy soil, respectively (Coastal Development Institute of Technology, 2019).

1.3.3 Column/element installation patterns and applications

One stroke of the mixing tools produces a column shape of stabilized soil. By the series of productions, arbitrary shape of stabilized soil ground can be constructed in

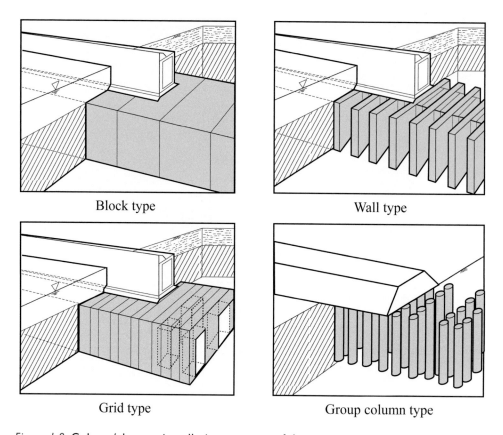

Figure 1.8 Column/element installation patterns of deep mixing improved ground.

the ground. When the deep mixing method is used as a solution for problems encountered on a construction project on soft ground, stabilized soil columns/elements are produced by a variety of column/element installation patterns to construct the block-, grid-, wall- or group column-type improved grounds, as shown in Figure 1.8.

Table 1.3 shows comparisons of the characteristics of the improved grounds. The block-, wall- and grid-type improved grounds are constructed by overlapping stabilized soil columns/elements. The block-type improved ground is the most stable against both the external and internal stabilities among all the improvement types. It may find application in case of breakwater, sea revetment or huge earth retaining structure which is subjected to large vertical and horizontal forces. The wall-type improved ground consisted of long walls and short walls, where the long walls are linked firmly by the short walls. This type is also stable against the external stability. The squeeze failure of unimproved soil between the long walls should be evaluated when the spacing of the long walls is relatively large (Terashi et al., 1983). The grid-type improved ground has almost the same function as the block-type improved ground with less stabilized soil volume, which can be applicable when the internal stability is less critical compared to the block type. When the instability is dominant to one direction, the wall type can effectively improve the stability. The wall-type or grid-type improved

Table 1.3 Characteristics of improvement types.

Type	Stability	Installation	Design Consideration
Block type	Large solid block resists external loads. Highly stable.	Takes longer time because all columns/elements are overlapped.	Design of size of block is in the same way as the gravity structures.
Wall type	Where all improved walls are linked firmly by short wall, high stability is obtained.	Requires precise operation of overlapping of long and short units.	Requires consideration of unimproved soil between walls. Wall spacing and depth of short wall affected by internal stability.
Grid type	Highly stable next to block type. Liquefaction prevention.	Installation sequences are complicated because lattice shape must be formed.	Requires design on three-dimensional internal stress.
Group column type	When horizontal loads are small, high stability is obtained.	Overlapping operation is not required.	Requires design on overall stability and on individual column as a pile foundation.
Tangent group column	When horizontal loads are small, high stability is obtained.	Precise operation is required to achieve tight and reliable contact of columns.	Requires design on overall stability and on internal stability of tangent columns.

ground may be selected in order to assure stability of embankment slope or foundation support for retaining structure. Recently, the grid-type improvement is frequently applied to liquefaction prevention, where shear deformation of unimproved soil between the grid is restricted so that the excess pore water pressure increase is controlled. When the major concern is the consolidation settlement of soft ground under embankment or a light weight structure, the group column-type improved ground will provide a good solution. The improvement area ratio is usually of the order of 30%–50%. The tangent group column-type improvement is a modified improvement pattern of the group column-type improvement, where stabilized soil columns are installed in contact with the adjacent columns without overlapping. As the improvement area ratio is usually of the order of 70%–80%, larger than that of the group column-type improvement, larger effects in bearing capacity and settlement reduction are expected than the group column-type improvement. This improvement has frequently applied to embankment slope and small building for increasing stability and bearing capacity, respectively.

1.4 SCOPE OF BOOK

The deep mixing method was developed in Japan and put into practice in the middle of 1970s. Many books are published to introduce ground survey, design, construction and construction control of the deep mixing method. However, few books about the quality control and assurance are published even if the quality of stabilized soil is greatly influenced by deep mixing equipment and mixing procedure. The text is aimed to provide the researchers and practitioners with the latest State of Practice of quality control and quality assurance of the deep mixing method based on the research studies done in Japan and experience accumulated by numerous projects since the mid-1970s to 2020 in Japan.

The organization of the current book is as follows:

Chapter 1 explained the deep mixing as a technique in the category of admixture stabilization. Also the diversity of deep mixing equipment and improvement patterns were shown, which includes the on-land deep mixing equipment and marine deep mixing equipment.

Chapter 2 focuses upon the quality control and assurance for deep mixing. The concept of QC/QA described in the chapter is generally applicable to all the admixture stabilization. However, quality control procedures during production differ for different mixing processes and also laboratory mix test program as a pre-production QA differs for different mixing processes. The current chapter focuses on the mechanical mixing by vertical shaft mixing tools with horizontal rotating circular mixing blade.

Chapter 3 discusses the technical issues on QC/QA of the method. The current QC/QA procedure may not always be possible to conduct the QC/QA due to several reasons, such as site conditions, and time and economical limitations. Some technical issues that may affect the evaluation of QC/QA are briefly explained.

Chapter 4 describes the stabilization of problematic soil. Some cohesive soils are so sticky that the soil and binder mixture adheres to and rotates with the mixing blade without efficient mixing, which causes less mixing and strength decrease of stabilized soil. Several countermeasures against such a soil are introduced in the chapter.

Chapter 5 shows applicability of *W/C* concept to the quality control during production. The quality of stabilized soil depends upon a number of factors but is in principle influenced by the combination of water, soil and binder. The water to cement ratio is one of the key factors to evaluate the strength of stabilized soil. The importance of water to cement ratio concept in the QC/QA and the future trend of QC/QA are discussed in the chapter.

REFERENCES

Broms, B.B. (1984) Stabilization of soil with lime columns. *Design Handbook*, 3rd Edition, Lime Column AB, Kungsbacka.

Broms, B.B. (1991) Stabilization of soil with lime columns. In: Fang, H.Y. (eds.) *Foundation Engineering Handbook*, pp. 833–855. Springer, Boston, MA.

Cement Deep Mixing Method Association (1999) *Cement Deep Mixing Method (CDM), Design and Construction Manual*. Cement Deep Mixing Method Association, 192p. (in Japanese).

Coastal Development Institute of Technology (2008) *Technical Manual of Pneumatic Flow Mixing Method, revised version*. Daikousha Publishers, 188p. (in Japanese).

Coastal Development Institute of Technology (2019) *Technical Manual of Deep mixing method for marine works, revised version*. Daikousha Publishers, 315p. (in Japanese).

Dry Jet Mixing Method Association (2006) *Dry Jet Mixing (DJM) Method Technical Manual*. Dry Jet Mixing Association, (in Japanese).

Ingles, O.G. and Metcalf, J.B. (1972) *Soil Stabilization, Principles and Practice*. Butterworth, Oxford.

JACSMAN Association (2011) *Technical data for JACSMAN, Ver. 6*. JACSMAN Association, 21p. (in Japanese).

Kitazume, M. (2011) Keynote lecture: Current practices in ground stabilization in Japan. Proceedings of the Korean Geotechnical Society Symposium on Recent Developments in Soft Ground Engineering for Overseas Projects, pp. 53–63.

Kitazume, M. and Terashi, M. (2013) *The Deep Mixing Method*. CRC Press, Taylor & Francis Group, Boca Raton, FL, 410p.

Miyoshi, A. and Hirayama, K. (1996) Test of solidified columns using a combined system of mechanical churning and jetting. *Proceedings of the 2nd International Conference on Ground Improvement Geosystems*, pp. 743–748.

Public Works Research Center (2004) *Technical manual on deep mixing method for on land works*. Public Works Research Center, 334p. (in Japanese).

Saitoh, S., Suzuki, Y. and Shirai, K. (1985) Hardening of soil improved by deep mixing method. *Proceedings of the 11th International Conference on Soil Mechanics and Foundation Engineering*. Vol. 3, pp. 1745–1748.

Terashi, M. and Kitazume, M. (2009) Keynote lecture: Current practice and future perspective of QA/QC for deep-mixed ground. *Proceedings of the International Symposium on Deep Mixing and Admixture Stabilization*, pp. 61–99.

Terashi, M. and Kitazume, M. (2011) QA/QC for deep-mixed ground: Current practice and future research needs. *Ground Improvement*, pp. 161–177.

Terashi, M. and Kitazume, M. (2015) Deep mixing - four decades of experience, research and development. *Proceedings of the Deep Mixing 2015*, San Francisco, June 2–5, pp. 781–200.

Terashi, M., Tanaka, H. and Kitazume, M. (1983) Extrusion failure of ground improved by the deep mixing method. *Proceedings of the 7th Asian Regional Conference on Soil Mechanics and Foundation Engineering*. Vol. 1, pp. 313–318.

The Building Center of Japan and the Center for Better Living (2018) *Design and Quality Control Guideline of Improved Ground for Building, 2018*. The Building Center of Japan and the Center for Better Living, 708p. (in Japanese).

The Japanese Society of Soil Mechanics and Foundation Engineering (1986) Ground improvement techniques - from soil investigation to design and construction. *The Japanese Society of Soil Mechanics and Foundation Engineering*, 330p.

Chapter 2

Quality control and assurance of deep mixing method

2.1 IMPORTANCE OF QUALITY CONTROL AND QUALITY ASSURANCE

The deep mixing (DM) improved ground should be constructed to assure the design criteria regarding the uniformity and strength of stabilized soil and the dimension and location of stabilized soil columns. In the design of DM improved ground, the size of improved zone, installation depth and pattern are determined based on the ground condition and external load condition and design parameters (geotechnical design) so that the improved ground can satisfy the performance criteria of the superstructure. In the process design, the construction control values to realize the quality of improved ground specified by the geotechnical design are specified. The specifications include not only the uniformity and strength of stabilized soil columns/elements but also the accuracy of production of the stabilized soil columns/elements in order to assure the location, depth, tight and reliable contact with a bearing layer and reliable overlap of columns/elements.

However, the quality of stabilized soil column/element such as uniformity and strength depends upon a number of factors including the type and condition of original soil, the type and amount of binder, and the execution equipment and process and the curing condition. The quality control and assurance, QC/QA, is essential to produce stabilized soil columns/elements that assure the design criteria and requirements on their quality and geometry. The quality control and quality assurance practice was originally established in Japan and Nordic countries in the 1970s and has been accepted worldwide for more than five decades. It comprises laboratory mix test, field-trial test, monitoring and controlling construction parameters during production and the verification after construction by measuring the engineering properties of stabilized soil either by unconfined compression tests on boring core samples or by sounding.

2.2 WORK FLOW OF DEEP MIXING PROJECT AND QC/QA

Quality assurance of the DM method to fulfill the requirements of geotechnical design cannot be achieved only by the process control during construction conducted by DM contractor, but it should involve relevant activities that are carried out prior to, during and after the construction by all the parties involved in the DM project.

DOI: 10.1201/9781003223054-2

Usually the site investigation of original ground, for example, is not considered as a part of QA, but it can be classified as one of the relevant activities. If the site investigation failed to identify existence of problematic layer, laboratory mix test would not be undertaken for the layer, which might result in the insufficient process design (including QC/QA methods/procedures) and would cause difficulty in interpretation of the field-trial stabilized soil columns/elements and/or verification test of production columns/elements.

The quality control and assurance (QC/QA) is one of the major concerns for clients and engineers who have less experience of the DM technique. The geotechnical design procedure in DM project is different depending on the application, and the construction control parameters and values are also different for different DM construction equipment. However, the overall work flow exemplified in Figure 2.1 (after Terashi, 2003) is common to the DM projects. Parties involved in DM project are project owner, design engineers, prime contractor, DM contractor and soil investigation and testing firm. In the figure, project owner's functions are shown in white frame, activities related to the geotechnical design are in slight gray frame, activities related to the process design and actual execution with QC are in light gray frame, and accumulated experience and database on locally available execution processes is shown in gray frame.

1. Performance requirements

    The project owner defines the functional and performance criteria and requirements of the superstructure, carries out site survey and site characterization studies, obtains information regarding the site-specific restrictions, and sometimes clarifies the purpose and requirements for ground improvement based on conceptual design.
2. Laboratory mix test

    The laboratory mix test may be carried out for relevant representative soil layers together with examination of experience of similar and accumulated database to study the applicability of the DM method to the project and to obtain design parameters.
3. Assumption and establishment of design parameters

    The geotechnical design regarding the DM is to determine the size, depth and column/element installation patterns and end-bearing condition (fixed-type or floating-type improvement) of the improved ground so that the DM improved ground can satisfy the functional and performance criteria and requirements of superstructure. Before making geotechnical design, designer should assume/establish the required quality of field stabilized soil (design parameters) and required level of accuracy of installation by taking into account the capability of DM construction equipment and personals available locally based on similar experience or local database. The design standard strength, q_{uck}, is one of the essential strength parameters in the design, while the compressive strength, σ_c, the tensile strength, σ_t, and the bending strength, σ_d, are derived from the q_{uck} (Kitazume and Terashi, 2013). It should be noted that the validity and accuracy of these assumptions cannot be confirmed before the execution but only after the field-trial test or the actual production. It is desirable to confirm them in the field-trial test to prevent any troubles in the actual production.

QC/QA of deep mixing method 17

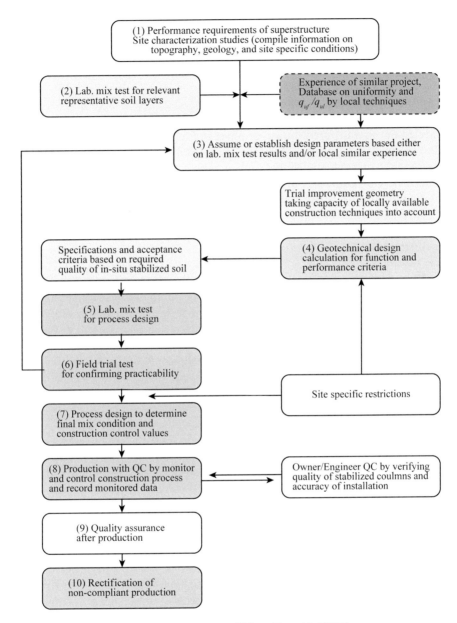

Figure 2.1 Workflow of deep mixing project (After Terashi, 2003).

4. Geotechnical design

The installation width, depth and pattern of the improved ground are determined based on the design parameters in the geotechnical design so that the improved ground can satisfy the functional and performance criteria and requirements of the superstructure. Many technical standards and manuals for the geotechnical design of DM improved ground are prepared for various improvement purposes and various improvement patterns (Coastal Development Institute of Technology, 2019; Dry Jet Mixing Method Association, 2006; The Building Center of Japan and the Center for Better Living, 2018; EuroSoilStab, 2002; Federal Highway Administration, 2013). The preparation of contract document including the acceptance criteria and verification procedure is, therefore, one of the important role of the geotechnical design.

5. Laboratory mix test for process design

   The laboratory mix test is an important pre-production QA which may be carried out in a different phase or phases of a project either for the geotechnical design and for the process design. It is carried out for the process design to determine the mixing conditions, such as type and amount of binder and chemical additives. The laboratory mix test is the responsibility of the owner/engineer if the DM contract work is awarded with detailed specifications, but is the responsibility of the DM contractor if the contract is awarded by performance basis. The mixing condition determined in the laboratory mix test is evaluated and confirmed in the field-trial test.

6. Field-trial test

   The field-trial test is also an important pre-production QA for DM project especially when no comparable experience is available. It is recommended to conduct field-trial test in advance in or adjacent to the construction site, in order to confirm the actual field strength and uniformity can be achieved in the real construction condition and determine the operational parameters such as the penetration and withdrawal speeds of mixing tool, *etc.* and final mixing condition for production. After the production, the quality of the field stabilized soil columns/elements should be verified in order to confirm the design quality of stabilized soil column/element, such as continuity, uniformity, strength, permeability or geometry can be achieved by the mixing conditions determined in the process design. If the design quality cannot be achieved by the mixing condition, and present equipment and process, they should be reviewed based on the field-trial test results and modified accordingly. An additional field test should be carried out until the mixing condition and DM equipment and procedure can satisfy the design criteria.

   The trial field test with injecting water instead of binder is a common practice in Japan to determine the criteria of process control value to confirm the end-bearing of column/element to the stiff layer in the case of the fixed-type improvement. The electric or hydraulic power consumption, torque and/or the penetration speed of mixing tool are measured during the field-trial test to establish the construction control criteria for the bottom treatment. The field-trial test for this purpose should be conducted in the vicinity of existing boring to compare with the known soil stratification and condition.

7. Process design

   The role of the process design is to determine the final mixing condition including the type and amount of binder and chemical additives, and the construction procedure, construction control factors and construction control values based on the laboratory mix test and the field-trial test. They should be determined appropriately in order to realize and assure the required quality of stabilized soil that the geotechnical design requires. A contract document is prepared in which the detail of specifications and tolerance are specified for each project; the binder factor, water to binder ratio, *w/c* of binder slurry, the penetration and withdrawal speeds of mixing tool, diameter, top and bottom depths of stabilized soil column/element, core recovery of boring core and the uniformity and strength of stabilized soil column/element, *etc.*

8. Production of stabilized column/element with QC

    Stabilized soil columns/elements must be produced to satisfy both the geometric layout and quality specified by the geotechnical design. Rig operator should locates, controls, monitors and records the geometric layout of each column/element (plan location, verticality, and top and bottom depths, *etc.*). When the termination depth is designated to ensure the tight and reliable contact to the underlying stiff layer (fixed-type improved ground), rig operator should carefully identify the termination depth according to the construction control criteria established in the process design. The construction control parameters during the production should be measured and recorded accordingly and submitted to the supervisor soon after the production of each column/element. This is an important QA during production because the quality of stabilized soil column/element can be preliminary evaluated by the record even before core boring the stabilized soil column/element.

9. Quality assurance after production

    After the production, the quality of the field stabilized soil columns/elements should be verified in advance to the construction of superstructure in order to confirm if the design quality, such as continuity, uniformity, strength, permeability or dimension can satisfy the design criteria. In Japan, the verification is usually carried out by means of observing and testing boring core samples of production columns/elements. The double and triple-tube core samplings are desirable for taking high quality of boring core samples (Cement Deep Mixing Method Association, 1999; Dry Jet Mixing Method Association, 2006; The Building Center of Japan and the Center for Better Living, 2018).

    The frequency of core boring is determined according to the total volume of stabilized soil and importance of superstructure. In the case of on-land works in Japan for example, three core borings are generally taken in the case where the total number of columns/elements is less than 500. When the total number exceeds 500, one additional core boring is taken for every further 250 columns/elements. In each core boring, boring core samples are taken throughout the depth in order to verify the uniformity and continuity of stabilized soil column/element by visual inspection. The engineering properties of the stabilized soil is usually determined based on unconfined compressive strength on boring core samples at 28 day curing.

10. Rectification of non-compliant column/element

    Any stabilized soil column/element is treated as a non-compliant column/element with the contract if any construction parameters or the quality of stabilized soil column/element is not satisfied with the contract specification. In this case, review and rectification of the non-compliant column/element are essentially. The contractor shall review the DM equipment and procedure based on the recorded production log data to investigate the cause for the non-compliant column/element and modify them accordingly. The influence of the non-compliant column/element to the performance and stability of the improved ground should be evaluated to determine the necessity and remediation measurement. It is frequently adopted to construct additional stabilized soil columns/elements close to the non-compliant column/element to compensate and reinforce the non-compliant column/element.

2.3 CURRENT PRACTICE OF QC/QA

2.3.1 Basic concept of laboratory, field and design standard strengths

Some issues regarding the current QC/QA practice are briefly introduced in this section, while the technical issues relating the QC/QA will be introduced in detail in Chapter 3.

The properties of stabilized soil is affected by many factors such as soil properties (natural water content, liquid limit, plastic limit, *pH*, organic matter content, grain size distribution, clay minerals, *etc.*), type and quantity of binder, mixing degree, and curing conditions, mixing condition and curing condition in a field are quite different from the laboratory mix test conditions, and therefore the properties of field stabilized soil are usually different from those of the laboratory stabilized soil. The field stabilized soil column/elements have relatively large strength variability even if the production is carried out with the established DM construction equipment and procedure and also with the best care. The unconfined compressive strength and deviation are schematically shown for the laboratory and field stabilized soils in Figure 2.2 (Kitazume and Terashi, 2013). Usually the field stabilized soil has smaller average strength and larger strength deviation than those of the laboratory stabilized soil. The relationship between the design standard strength, q_{uck}, and the average of field strength, q_{uf} can be formulated by incorporating the strength deviation as Equation (2.1), if the strengths of the laboratory and field stabilized soils are assumed to have a normal distribution.

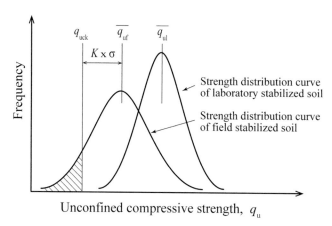

Figure 2.2 Strengths of laboratory and field stabilized soils (Kitazume and Terashi, 2013).

$$q_{uck} = \overline{q_{uf}} - K \times \sigma = \overline{q_{uf}} \times (1 - K \times COV) \tag{2.1}$$

$$\overline{q_{uf}} = \lambda \times \overline{q_{ul}}$$

where
  COV: coefficient of variation
  K: coefficient
  q_{uck}: design standard strength (kN/m^2)
  $\overline{q_{uf}}$: average unconfined compressive strength of field stabilized soil (kN/m^2)
  $\overline{q_{ul}}$: average unconfined compressive strength of laboratory stabilized soil (kN/m^2)
  σ: standard deviation (kN/m^2)
  λ: strength ratio of q_{uf}/q_{ul}

The magnitude of the parameter K can be determined by the probability, which is defined as the frequency of the strength lower than the q_{uck}, as a portion drawn with a hatch in Figure 2.2. The magnitude of the probability should be determined in the geotechnical design by considering the type, size and importance of superstructure and also the capability and performance of DM equipment and personnel. The relationship between the K and probability is shown in Table 2.1 assuming that the distribution of the field strength has a normal distribution.

2.3.2 Process design

The role of the process design is to determine the final mixing condition such as the type and amount of binder and chemical additives, and the construction procedure, construction control parameters and construction control values based on the laboratory mix test and the field-trial test. They should be determined appropriately in order to realize and assure the required quality of field stabilized soil that are required in the geotechnical design.

The laboratory mix test is an important pre-production QA for the process design. It is carried out to determine the mixing conditions, such as type and amount of binder and chemical additives. It is found that nationwide (or regional) official standards or guidelines of laboratory mix test are scarce, but several kinds of testing procedures have been adopted in each region and organizations (Terashi and Kitazume, 2009). These testing procedures can be derived basically from the two concepts: (a) reproduce of field mixing condition, and (b) reproduce of ideal mixing condition, which will be explained in detail in Section 3.1, Chapter 3. In this section, the Japanese process design that is derived based on Concept B is introduced.

Table 2.1 Relationship between K and probability.

Parameter, K	0.5	1.0	1.645	2.0	3.0
Probability (%)	30.9	15.9	5.0	2.3	0.13

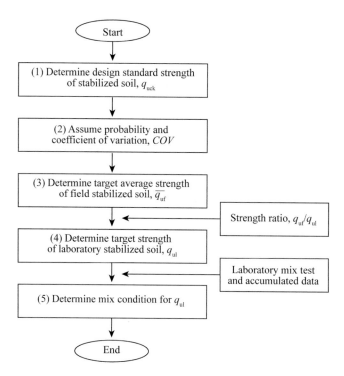

Figure 2.3 Flow of process design of stabilized soil (After Coastal Development Institute of Technology, 2008).

2.3.2.1 Flow of mixing design and process design

Figure 2.3 shows the flow of the process design of stabilized soil that is frequently adopted in Japan (after Coastal Development Institute of Technology, 2008).

1. At first, the design standard strength of stabilized soil, q_{uck}, is determined in the geotechnical design for assuring the design criteria and performance of the superstructure and DM improved ground.
2. Based on the accumulated data base, the target average field strength, q_{uf}, is determined by Equation (2.1) together with the assumed magnitude of probability and the coefficient of variation of field stabilized soil strength. The relationship between the K and the probability is already shown in Table 2.1. The magnitude of the probability should be determined as an engineering judgments by taking into account the function, type and size of superstructure, the purpose of DM ground, the capacity and ability of DM equipment and operator, and so on. However, the K of 1.0 (probability of about 16%) has been often adopted in Japan.
3. The target average laboratory strength, q_{ul}, is determined by considering the strength ratio, q_{uf}/q_{ul}. There are a lot of database and proposal on the strength ratio, q_{uf}/q_{ul}. However, the strength ratio, q_{uf}/q_{ul}, is obviously influenced not only by the soil and binder types and DM equipment and procedure but also the

laboratory mix test procedure. An appropriate value of the strength ratio should be adopted from the data base depending on the field condition and the laboratory mix test condition. The assumed strength ratio should be confirmed in the field-trial test.

4. The laboratory mix test should be conducted on soil samples retrieved from all the soil layers to be stabilized, in order to determine a suitable type and amount of binder to achieve the design strength. In the laboratory tests, a productive and high repeatable specimen are produced to simulate an ideal mixing condition and specimen. The amount of binder is usually changed about three stages to obtain the target amount of binder by interpolating the test results. Ordinary Portland cement and blast furnace slag cement type B (containing 30%–60% slag) are usually used as a binder both in the wet and dry methods in Japan. Dozens of special binders are also available on the Japanese market for organic soils and extremely soft soil with high water content (Japan Cement Association, 2012) and they are used for the laboratory mix test when required.

Chemical additive is sometimes mixed together with binder for specific soils and special purposes, such as to keep the fluidity of binder slurry and fluidity of soil and binder mixture high and keep the short term strength low. Several examples on applicability of chemical additives will be introduced in Section 4.2.3 in Chapter 4 (Nozu et al., 2012, 2015; Aoyama et al., 2002; Hirano et al., 2015; Mizutani and Makiuchi, 2003). Their type and amount are also determined in the laboratory mix test.

2.3.2.2 Mixing condition in laboratory and field

The strength ratio of laboratory strength to field strength of stabilized soil is one of the essential parameters for the process design. The strength ratio is much dependent upon the mixing condition and curing condition. Besides these conditions, the mixing degree in laboratory and field caused by the mixing techniques should be emphasized here. Figure 2.4 schematically illustrates the relationship between the strength of stabilized soil and amount of binder for mechanical mixing technique.

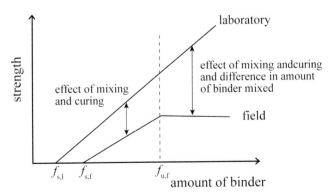

Figure 2.4 Relationship between strength of stabilized soil and amount of binder.

Figure 2.5 Electric mixer and mixing blade for the laboratory mix test.

In the laboratory mix test, the soil and binder are put in a mixing bowl together and mixed with a mixing panel, as shown in Figure 2.5. The soil and binder mixture do not come out of the bowl during the mixing and they are finally mixed well to the uniform mixture. The strength of the stabilized soil produced in the laboratory is increased monotonically with the increase of the amount of binder when the binder factor exceeds a certain value (threshold value). The threshold value, $f_{s,l}$, is dependent on the type and condition of soil and the type of mixer, *etc.*, but it is around $50\,kg/m^3$ according to the past experience.

The field strength is also increased with the increase of the amount of binder. However, the threshold value for the field stabilized soil, $f_{s,f}$, is depend on the DM equipment and procedure as well as the type and condition of original soil. It is usually larger than that for the laboratory stabilized soil and around $80–100\,kg/m^3$ according to the case histories. The field strength increasing ratio with the amount of binder is usually smaller than the laboratory strength, which is due to the difference in the mixing and curing conditions. When the amount of binder exceeds a certain level, a part of binder exceeding the level flows out to ground surface without mixing with the in-situ soil, which causes that the field strength remains almost constant. The level, a sort of upper limit of binder, $f_{u,f}$, is depend on the ground and mixing conditions and is around $250–300\,kg/m^3$ for the wet method according to the experience. According to the above, the amount of binder should be determined within the threshold, $f_{s,f}$, and the upper limit, $f_{u,f}$. If not, it is necessary to use any special binder or cement to achieve the design strength or to reduce the design standard strength to the achievable level in the geotechnical design.

2.3.2.3 Tips of laboratory mix test

It is often seen that the laboratory mix test is carried out by changing the amount of binder alone without changing the water content of soil and the water to binder ratio, w/c, of binder slurry. It may be sufficient as far as particular soil condition, mixing condition and DM equipment condition are concerned. However, it can be seen in many projects that the water content of original soil is not uniform but varies along the depth and also site by site even in the same soil type. It is not desirable, but some amount of water is sometimes injected to expect smooth penetration of mixing tool when encountering a hard layer. It can't always facilitate the penetration of mixing tool but always causes the strength decrease and ground heaving increase. It is also usual that the water to binder ratio, w/c, of binder slurry is different for each DM equipment depend on the capacity of slurry pump and the layout and dimension of pipeline facility. The binder slurry with low w/c and low fluidity requires high pumping capacity and construction ability, but it is preferable for large strength gain and reduction of ground deformation and heaving due to production of columns/elements. The w/c of binder slurry assumed in the process design is not always the same as the DM construction equipment of awarded contractor. In this case, the mixing condition determined in the laboratory mix test is not applied in the field. It is desirable to carry out the laboratory mix test by changing not only the amount of binder but also the water content of soil and the w/c ratio of binder slurry widely in order to correspond to the equipment type and field condition.

Figure 2.6 illustrates a relationship between the amount of binder and the strength of stabilized soil, in which three curves are shown as an example for three different water contents of soil; w_1 for natural water content, w_2 for little larger water content than the natural water content, and w_3 for higher water content than the natural water content. Three different binder factors, F_1, F_2 and F_3, can be obtained for the target laboratory strength, q_{ul}, for each water content. The amount of binder is determined appropriately to obtain the target laboratory strength, q_{ul}, and target field strength, q_{uf}, by considering the measured/expected variation of water content of original soil, the possibility of water injection and the expected w/c ratio of binder slurry. As described before, some amount of water is injected when encountered a hard layer in some cases, which causes the increase of in-situ water content. In this case, the amount of binder should be increased to assure the design strength. The target amount of binder can be obtained as F_2 in Figure 2.6a as an example if the laboratory mix test was carried out for the "water content 2."

Figure 2.6b illustrates the relationship between the water to binder ratio, W/C and the unconfined compressive strength. The W/C is defined as the ratio of the total weight of water containing binder slurry and soil to the weight of soil, as Equation (2.2).

$$\begin{aligned} W/C &= (W_{ws} + W_{wc} + W_{ww})/W_c \\ &= (w/aw + w/c) + W_{ww}/W_c \end{aligned} \qquad (2.2)$$

where
  aw: binder content (%)
  w: water content of original soil (w)
  w/c: water to binder ratio of binder slurry

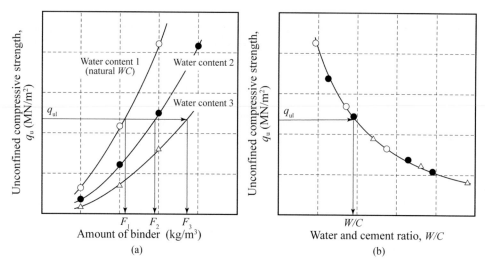

Figure 2.6 Determination of amount of binder. (a) Amount of binder and strength. (b) W/C ratio and strength.

W/C: water to binder ratio of stabilized soil
W_c: weight of binder (kg)
W_{wc}: weight of water of binder slurry (kg)
W_{ws}: weight of water contained in soil (kg)
W_{ww}: weight of injected water (kg)

An unique line can be provided experimentally in the W/C and strength diagraph for each soil type and binder type irrespective of the water content of soil and the w/c of binder slurry. The figure will be helpful for the case where the water content of in-situ soil and/or the water to binder ratio of binder slurry, w/c, is different from the originally estimated values.

It is desirable to measure the water content of the piece of stabilized soil sample after the unconfined compression test for obtaining the relationship with the q_{ul}. This relationship can provide useful and helpful inspiring in the quality assurance especially when discussing about the non-compliant stabilized soil column/element.

2.3.3 Selection of deep mixing equipment

As shown in Figures 1.4–1.7, the size, shape and layout of mixing blade and the position of injection nozzle greatly varies according to DM equipment, which gives large influence on the mixing performance and uniformity and strength of stabilized soil column/element. For assuring sufficient mixing with binder and soil, it is essential to prevent the entrained mixing phenomenon as much as possible, in which disturbed soil adheres to and rotates with the mixing blade without efficient mixing of soil and binder, as shown in Figure 2.7. For a single-shaft equipment, a free blade, an extra blade about 100 mm longer than the diameter of mixing blade, is installed close to

Figure 2.7 Example of entrained mixing phenomenon.

one of the mixing blades to prevent the entrained mixing phenomenon. A multi-shaft equipment has a bracing plate to keep the distance of mixing shafts (see Figure 1.5). The plate is also expected to function to increase mixing degree by preventing the entrained mixing phenomenon. The mixing shafts of the multi-shaft equipment rotate in the opposite direction each other, which increases the degree of mixing and also mitigates the oscillation of DM equipment itself.

The diameter, shape, angle, number and layout of the mixing blades influence the mixing degree. A lot of research efforts and filed tests were performed to find suitable shape, number and layout of the mixing blades (see Section 3.4 in Chapter 3). The suitable equipment with efficient mixing tool should be selected according to the past experiences and its efficiency and performance should be confirmed in the field-trial test for the specific ground condition.

2.3.4 Field-trial test

The field-trial test is also an important pre-production QA for DM project especially when no comparable experience is available. It is recommended to conduct field-trial test in advance in or adjacent to the construction site, in order to evaluate the applicability and reliability of the DM equipment and procedure and to obtain suitable operational parameters and final mix design for production to assure the design criteria, uniformity and strength of stabilized soil and dimension of the stabilized soil columns/elements in the real construction condition. For the purposes, it is advised to carry out the field-trial changing the amount of binder and also construction parameters such as the penetration and withdrawal speeds of mixing tool and the blade rotation number. In the case where the quality (strength and *COV, etc.*) of stabilized soil columns/elements specified the design cannot be achieved by the DM equipment and procedure, it is necessary to remodel the equipment and procedure or modify the design of improved ground according to the field-trial test results.

The water to binder ratio, *w/c*, of binder slurry depends on the DM equipment. A low *w/c* ratio is preferable for strength gain and reduction of ground heaving, but not for construction ability due to its low fluidity. According to the Japanese experiences, the *w/c* ratio is adopted to 60–100% in many wet method equipment (Cement Deep Mixing Method Association, 1999).

2.3.5 Quality control during production

2.3.5.1 Construction procedure

There are two construction procedures basically depending on the injection sequence of binder (Figure 2.8): (a) injecting binder during the penetration of mixing tool (penetration injection method) and (b) injecting binder during the withdrawal of mixing tool (withdrawal injection method). Many Japanese DM equipment install the two injection nozzles for the two methods, as shown in Figure 2.9. The injection nozzle installed close to the lowermost mixing blades is used for the penetration injection, but one installed close to the uppermost mixing blade is used for the withdrawal injection.

1. penetration injection method

    Figure 2.6a shows a basic execution procedure of the penetration injection method. After setting DM equipment at the prescribed position, the mixing tool is penetrated into a ground while rotating the mixing blades continuously. The excavation blades installed at the bottom end of the mixing shaft cut and disturb the in-situ soil to reduce its strength so that the mixing tool can be easily penetrated in a ground by their self-weight. At the same time, the binder slurry is injected in the ground from the bottom injection nozzle. The flow rate of binder slurry is kept constant while the penetration speed is controlled constantly so as to assure the designed amount of binder should be injected and mixed with the soil *in-situ*.

    The stabilized soil columns/elements should be produced to the design depth in the floating-type improvement. In the case of the fixed-type improvement, the stabilized soil columns/elements should reach a stiff layer sufficiently. Since the depth of the stiff layer is not always constant but undulates in many cases, the stabilized soil column/element should be produced to the stiff layer by detecting the depth of the layer for each column/element. In practical execution, the penetration speed, driving torque and rotation speed of mixing blades are continuously monitored and compared with the construction control values established in the field-trial test to detect if the mixing tool reach the stiff layer. When the mixing tool reaches the stiff layer, the mixing tool stay there for several minutes with

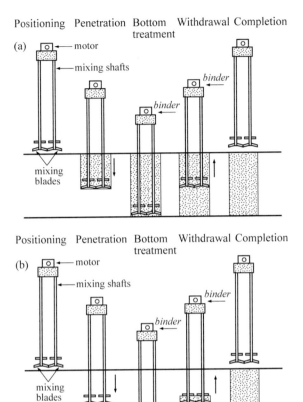

Figure 2.8 Execution procedure of deep mixing method. (a) Penetration injection method. (b) Withdrawal injection method.

    continuous binder injection and mixing or go up and down about one meter with continuous binder injection and mixing to assure the tight and reliable contact of the column/element with the stiff layer (bottom treatment). In the withdrawal stage, the mixing blade is rotating reversibly to mix the in-situ soil and binder again.

2. withdrawal injection method

    Figure 2.6b shows a basic execution procedure of the withdrawal injection method. After setting DM equipment at the prescribed position, the mixing tools are penetrated into a ground while rotating the mixing blades continuously, as similar manner to the penetration injection method. The excavation blades installed at the bottom end of the mixing shaft cut and disturb the in-situ soil to reduce its strength so that the mixing tool can be easily penetrated.

    The stabilized soil columns/element should be produced to the design depth in the floating-type improvement. In the case of the fixed-type improvement, the stabilized soil columns/elements should reach a stiff layer sufficiently. Since the depth of the stiff layer is not constant but undulated in many cases, the stabilized

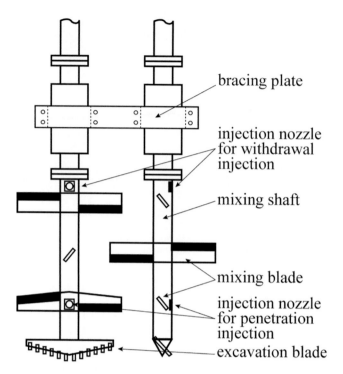

Figure 2.9 Basic layout of mixing tools and injection nozzles.

soil column/element should be produced to the stiff layer by detecting the depth of the layer for each column/element, as similar manner to the penetration injection method. When the mixing tool reaches the stiff layer, the mixing tool stay there for several minutes with continuous binder injection from the bottom injection nozzle and mixing or go up and down about one meter with continuous binder injection from the bottom injection nozzle and mixing to assure the tight and reliable contact of the column/element with the stiff layer (bottom treatment).

In the withdrawal stage, the mixing blades are rotating reversibly with injecting binder in the ground from the top injection nozzle. The flow rate of binder is kept constant while the withdrawal speed is controlled constantly so as to assure the designed amount of binder should be injected and mixed with the in-situ soil. It is necessary to overlap the depths of binder injection from the bottom and top injection nozzles for producing assuring the continuous stabilized soil column/element.

2.3.5.2 Overlap columns/elements

One series of penetration and withdrawal of mixing tool produces a column shape of stabilized soil in a ground. By the series of productions, arbitrary shape of stabilized soil ground can be constructed. When the deep mixing method is used as a solution for problems encountered on a construction project on soft ground, stabilized soil columns/elements are installed by a variety of column/element production patterns to construct block-, grid-, wall- or group column-type improved grounds, as already shown in Figure 1.8. In the overlapping procedure, a part of the previously produced stabilized soil column is scraped by the mixing blades and overlapped with a new column. It is desirable to complete the overlapping before the strength of previously produced stabilized soil column becomes large in order to achieve the tight connection.

Figure 2.10 shows the direct shear test results on the effect of overlapping time on the shear strength of the overlap joint face of stabilized soil (Yoshida, 1996). In the test, the stabilized soil elements were produced at several prescribed time intervals where the column with 1.0 m in diameter was produced with a special cement with a *w/c* ratio of 100% by a two mixing shafts equipment. The top surface of the elements was exposed to take boring core samples at the overlap joint portion as well as at the center portion of the column, and the boring core samples was trimmed to prepare the specimen for the direct shear test to measure their shear strength. The figure clearly shows that the shear strength at the overlap joint portion is about 60%–75% of that of the column center as long as the overlapping was completed within 4 days. However, the shear strength of the overlap joint portion is decreased to almost zero when the overlapping was carried out at 6 days after the previous production column.

Figure 2.11 shows an example of the field test at Kure Port to investigate the possibility of overlapping columns/elements changing the interval of overlapping (Yamada and Furuhashi, 1985). Figure 2.11a and b shows the hydraulics for driving the mixing tool and hanging load of mixing tool along the depth. The hydraulics is increased, and the hanging load is decreased with the depth. As shown in Figure 2.11a, the hydraulics in the cases of the interval is less than 24 hours are almost the same, but it becomes

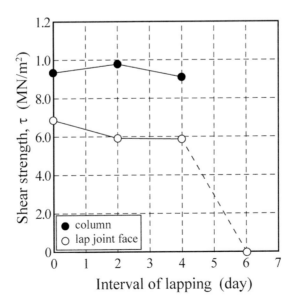

Figure 2.10 Shear strengths of stabilized soil at overlap joint and center of column (Yoshida, 1996).

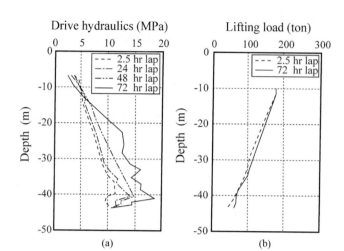

Figure 2.11 Drive hydraulics and hanging load distributions along depth (Yamada and Furuhashi, 1985). (a) Hydraulics for driving mixing tool. (b) Hanging load of mixing tool.

large when exceeding about 24 hours. According to the field observation, it was quite hard to scrape the previously produced column/element and overlap with a new column/element after 48 hours. A similar field test was carried out at Yokohama Port (Nakamura, 1977a, b).

According to the laboratory and field tests, it is specified to complete the overlapping within 24 hours from the previously produced columns/elements in order to achieve the tight connection (Cement Deep Mixing Method Association, 1999). The width of overlapping is also specified as at least 20 cm. However, in the case of a large-scale project, it becomes difficult to complete the overlapping within the specified time especially in the block-type improved ground, where all the columns/elements should be overlapped each other. In addition, the overlapping cannot be completed within the specific time in some cases due to bad weather and/or equipment trouble, which causes un-overlapping portion, 'cold joint,' in the improved ground. The cold joint may cause less stable and large deformation of the improved ground depending on its location and direction in the improved ground. In the case when the cold joint can't be avoided by any reasons, its location and direction should be determined to minimize its adverse influence to the stability and performance of improved ground (Kitazume and Imai, 2021).

2.3.5.3 Operational parameters

Figure 2.12 shows the operational parameters for the CDM method (Japanese wet method) that contains the quality and the geometric layout of stabilized soil column/element (after Cement Deep Mixing Method Association, 1999).

1. quality of stabilized soil column/element

    The water to binder ratio, *w/c*, of binder slurry (for wet method), volume of binder and mixing are controlled in the quality of stabilized soil. The weights

QC/QA of deep mixing method   33

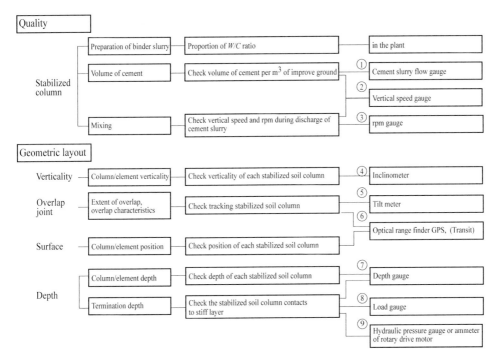

Figure 2.12 Operation monitoring for CDM method on-land works (After Cement Deep Mixing Method Association, 1999).

of binder and water are controlled to produce uniform binder slurry with the prescribed *w/c* ratio. The binder slurry should be used up within about 1 hour after preparation to prevent the setting of binder before injecting into the soil. The injected volume of binder is measured and controlled by the flow gauge. The degree of mixing is controlled by the vertical speed and rotation speed of mixing blade.

The vertical speed of the mixing tool and rotation speed of mixing blade are usually configured for the deep mixing method, as shown in Table 2.2 (Cement

Table 2.2 Typical execution control values of CDM method (Cement Deep Mixing Method Association, 1999).

Type	Penetration injection method	Withdrawal injection method
Mixing shaft		
Penetration speed (m/min)	1.0	1.0
Withdrawal speed (m/min)	1.0	0.7
Mixing blade rotation speed		
Penetration (rpm)	20	20
Withdrawal (rpm)	40	40
Blade rotation number (N/m)	360	350

Deep Mixing Method Association, 1999). They are slightly different depend on the injection method in order to achieve the same mixing degree. In the practical execution, the penetration and withdrawal speeds of mixing tool are controlled to the prescribed speed by sending out the suspended wire. The flow rate of binder slurry is also controlled constantly. During production of column/element, construction control values are controlled, monitored and displayed in the control room in the plant and/or in the cab of the deep mixing equipment for the plant operator and rig operator to adjust the execution procedure when necessary.

The mixing degree mostly depends on the rotation speed and penetration and withdrawal speeds of the mixing tool. In Japan, an index named blade rotation number, T has been introduced to evaluate the mixing degree. This number means the total number of mixing blade passes during 1 m mixing blade movement and is defined by Equations (2.3a) and (2.3b) for the penetration injection method and withdrawal injection method, respectively. According to the previous research studies and field experiences, the average strength of stabilized soil is increased with the increase of the blade rotation number and the coefficient of variation is decreased irrespective of soil type, type of deep mixing equipment and mixing condition. According to the accumulated research studies and investigations, the minimum blade rotation number should be around 270 for the dry method and 350 for the wet method to assure sufficient mixing degree for Japanese wet and dry methods, CDM and DJM methods (Cement Deep Mixing Method Association, 1999; Coastal Development Institute of Technology, 2019; Dry Jet Mixing Method Association, 2006; Public Works Research Center, 2004).

For penetration injection method,

$$T = \sum M \times \left(\frac{N_d}{V_d} + \frac{N_u}{V_u} \right) \quad (2.3a)$$

For withdrawal injection method,

$$T = \sum M \times \left(\frac{N_u}{V_u} \right) \quad (2.3b)$$

where

N_d: rotation speed of mixing blades during penetration (N/min.)
N_u: rotation speed of mixing blades during withdrawal (N/min.)
T: blade rotation number (N/m)
V_d: penetration speed of mixing blades (m/min.)
V_u: withdrawal speed of mixing blades (m/min.)
ΣM: total number of mixing blades

2. Geometric layout of stabilized soil columns/elements

The verticality, overlap joint, position and depth of stabilized soil column/element are controlled in the quality control of the geometric layout. The verticality and overlap joint are evaluated by the measurements of inclinometer. The position of the column/element is controlled by the GPS system. The top and bottom of column/element are controlled by gauges and also hydraulic pressure gauges. In the case of the block-, wall- and grid-type improvements (Figure 1.8), the tight overlap joint of the stabilized soil columns/elements are essential to maintain the shape of improved ground and to assure its performance. It is vital to place the mixing tool at the design position and penetrate it into the ground vertically for assuring the tight overlapping with previously produced stabilized soil columns/elements. The tolerances of the position and verticality of the column/element are specified at 0.2 m and 1/200 to 1/100, respectively, in Japan in order to ensure the overlapped width of 0.2 m and/or more. In the case of marine works, deep mixing barge shall be anchored suitably to maintain the position and verticality that may be easily disturbed by wave and wind.

3. Bottom treatment

    At the bottom of stabilized soil column/element, the blade rotation number is not automatically guaranteed for well mixing. A careful bottom mixing process by repeating penetration and withdrawal, while injecting binder is usually conducted to attain the sufficient level of mixing. When the quality of bottom end of column/element is critical such as the fixed-type improved ground, the quality at the bottom of column/element should be confirmed during the field-trial test.

2.3.5.4 Example of construction procedure

As described before, the operational parameters are controlled and recorded during the production. Figure 2.13 shows an example of production log data in wet method, in which an uniform soft ground stratified from −10 to −30.0 m was stabilized to construct the fixed-type improved ground by the wet method with the withdrawal injection.

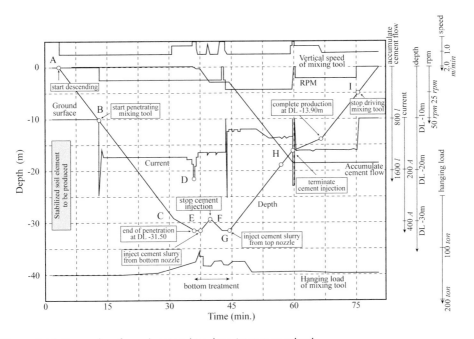

Figure 2.13 Example of production log data in wet method.

In the figure, the position of mixing shaft, penetration and withdrawal speeds of mixing shaft, rotation speed of mixing blade, electric current for driving mixing blade, lifting load of mixing blades and accumulated cement volume are plotted. The log data in the figure are briefly explained as follows:

At time A: The mixing tool was set at the designed position and the mixing tool was descended to a ground surface at a speed of 1.0 m/min, at that time the lifting load indicates 150 tons corresponding the self-weight of the mixing shaft and mixing blade.

At time B: When the mixing tool reached the ground surface at −10 m depth, the mixing blades were started rotating at a rotation speed of 23 rpm and the electric current for driving the mixing blade is jumped up to about 300 A at once and then to down to a constant value of about 150 A soon after. The mixing tool was penetrated in the ground at 1.0 m/min.

At time C: As approaching the stiff layer, the penetration speed was reduced to 0.3 m/min, while the rotation speed of mixing blades was kept constant of 23 rpm. The hanging load was gradually decreased as penetration due to the upward loads of the ground resistance and buoyancy.

At time D: The mixing tool reached the stiff layer at −30.0 m as the quick increase in the electric current and quick decrease in the hanging load.

At times E, F and G: After reaching the stiff layer, the cement slurry was injected from the bottom injection nozzle to start the bottom treatment. The mixing tool was lifted up to the depth of −29.10 m once with continuous injecting the cement slurry from top injection nozzle (time F). The mixing tool was penetrated again to −31.50 m without cement injection (time G). The mixing blades were continuously rotated at 23 rpm during the bottom treatment to assure the tight and reliable contact of the stabilized soil column/element to the stiff layer.

At time G: After the bottom treatment, the mixing tool was lifted up at 1.0 m/min to −15.0 m, while the cement slurry was continuously injected from the top injection nozzle at 823 l/min, and the mixing blades were continuously rotated at 45 rpm to mix the in-situ soil and cement. The electric current was 50–70 A and smaller than that in the penetration stage due to disturbing and softening the soil in the penetration stage. The hanging load was increased again due to the loss of upward loads of the ground resistance and buoyancy.

At time H: The cement injection was terminated at −16.20 m and then mixing tool was lifted up to −13.90 m at a rotation speed of 23 rpm.

At time I: The rotation of mixing blade was terminated at a depth of −5.0 m to complete the production of element. According to the monitored data, the 16.50 m length stabilized soil element was produced from a depth of −15.0 to −31.50 m.

2.3.6 Quality control throughout construction period

2.3.6.1 Material management

Quality control in the deep mixing method also includes the binder storage, binder or binder slurry preparation and control of the mixing process. Storage and mixing

Figure 2.14 Mud balance apparatus (https://yahoo.jp/Cxwb4Y).

of binder, additives and water are controlled, monitored and recorded in the plant throughout the project. The amount of binder supply is recorded by the shipping ticket, and the quality of binder is usually evaluated by the mill sheet and sometimes by the compression tests on the binder slurry samples. The water to binder ratio of binder slurry, w/c, is measured by the mud balance test at the prescribed time interval, as shown in Figure 2.14. All the data should be recorded throughout the project and submitted to the client for the inspection.

2.3.6.2 Modification of construction control values

Depending on the contract scheme, the construction control values can be modified accordingly during the construction period based on the examination of the previously produced columns/elements.

2.3.6.3 Damage of mixing tool

The mixing tool is sometimes defected, broken and bended by obstacles and large cobbles in a ground, as shown in Figure 2.15. It is necessary to inspect the mixing tool after each column/element production to confirm if all mixing tool is intact without any bending and missing blade. A bending and missing blade shall be replaced immediately before the next production of columns/elements.

2.3.6.4 Lateral displacement and ground heaving

As a result of injecting binder into a ground, the surrounding soil is displaced horizontally and ground surface heaves to some extent. Figure 2.16 shows an example of the measured lateral displacement at the ground surface in on-land works in reference to

Figure 2.15 Damaged mixing tool. (a) Bended mixing blade. (b) Defected mixing blade.

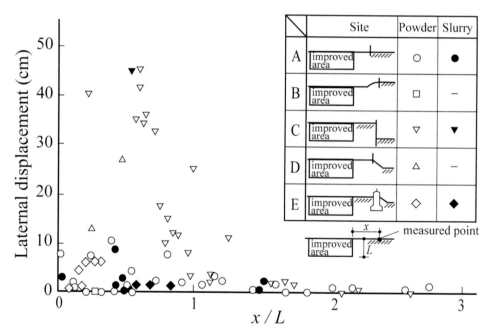

Figure 2.16 Lateral displacement of surrounding ground during improvement (Mizuno et al., 1986).

the local topography (Mizuno et al., 1986). Although the amount of ground movement is relatively small in on-land works compared with that of marine works (Cement Deep Mixing Method Association, 1999), the ground near the excavation and slope moves horizontally 10–40 cm. The amount of ground displacement and ground heaving depend upon the improvement area ratio and volume of binder slurry. According to the accumulated field experiences, the volume of heaved ground is about 60%–80% of the volume of injected binder slurry. The ground movement should be carefully measured

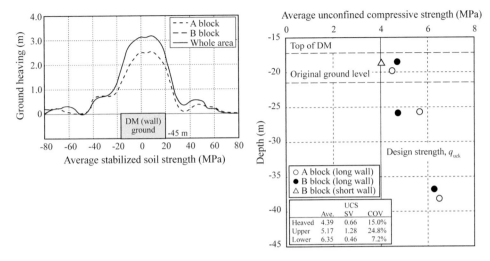

Figure 2.17 Ground heaving and strength profile in marine work (Yoshino et al., 2002). (a) Heaved ground level profile. (b) Strength distribution along depth.

and controlled to prevent adverse influence to the surrounding structures. During the production in on-land works, the heaved ground is excavated and removed by backhoe to prevent any adverse influence to the execution. It is desirable to reduce the w/c ratio and volume of binder slurry for reducing the ground displacement and heaving.

Figure 2.17 shows an example of ground heaving in the marine work (Yoshino et al., 2002). There is an alluvial clay layer of 25 m thickness at the seabed of −20 m, where its water content is 80%–120% in the upper layer (to −12 m) and 60%–80% in the lower layer (from −12 m). A sand mat of 1 m thickness was constructed on the ground surface before the production of stabilized soil element. Then, the ground was stabilized with blast furnace slag cement type B of the cement factor of 150 kg/m^3 to construct the wall-type improved ground. The cement factor was increased to 180 kg/m^3 in the heaved ground to achieve the design standard strength, q_{uck}, of 4 MPa.

Figure 2.17a, the heaved ground level profile at the two improvement blocks, shows a large ground heaving ranging from 1.3 to 3.5 m in the improved areas. The volume of heaved ground in the improved area is about 66% of the volume of injected cement slurry, which is the same order of the previous case record (Cement Deep Mixing Method Association, 1999).

Figure 2.17b shows the unconfined compressive strength distribution along the depth. The average strength at the alluvial layer is 51.7 and 63.5 MPa for the upper and lower layers, respectively, which is larger than the design standard strength of 4.0 MPa. The average strength at the heaved ground is smaller than that of the alluvial layer even if the cement factor was increased from 150 to 180 kg/m^3 but larger than the design standard strength with relatively small variation.

In order to reduce the ground displacement and heaving from the view point of deep mixing equipment, a new deep mixing construction equipment, the CDM-LODIC method (low displacement and control method) (Cement Deep Mixing Method Association, 2006; Kamimura et al., 2009) was developed and successfully employed in a number of on-land works, as shown in Figure 2.18. The CDM-LODIC method, a

Figure 2.18 Mixing tool of CDM-LODIC method (By the courtesy of Cement Deep Mixing Method Association).

variation of the CDM methods for on-land work, was developed in 1985 for minimizing the ground displacement and heaving during the execution.

Soil equivalent to the amount of mixing shaft and mixing blades and amount of injected binder slurry is removed before/during the binder slurry injection. Figure 2.19 shows the horizontal displacement measured by inclinometer during the production by an ordinary CDM equipment and the CDM-LODIC equipment (Kamimura et al., 2009). It is obvious that the CDM-LODIC method can reduce the horizontal displacement considerably. According to the accumulated data, it is found that the horizontal displacement caused by the CDM-LODIC equipment is less than 20 mm and quite smaller than that by the ordinary CDM equipment.

2.3.7 Report

Reporting the recorded construction control parameters is also an important QA throughout construction period. The production log data as well as the position, verticality and top and bottom depths of each stabilized soil column/element should be recorded and submitted to the client for the inspection. Table 2.3 shows an example of item to be reported.

QC/QA of deep mixing method 41

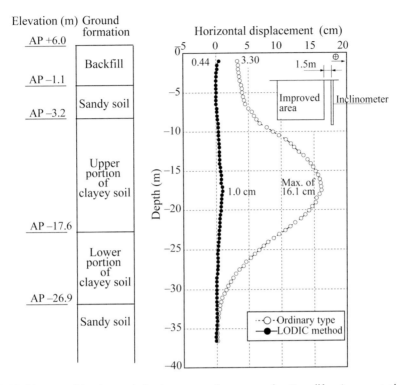

Figure 2.19 Measured horizontal displacement due to production (Kamimura et al., 2009).

Table 2.3 List of report document.

Item	Report
Delivery of binder	
Amount	Shipping ticket
Quality	Mill test certificate
Preparation of binder slurry	
Amounts of binder to water, w/c ratio	Weight measurements
Density	Mud balance test
Production of column/element	
Quality	Production log data (e.g. Figure 2.11)
Geometric layout	Position measurement and production log data
Verticality, column/element length	
Remarks	
Damage of mixing tool	Visual inspection
Ground heaving	Ground survey
Environmental impact	Noise, vibration, water pollution measurements, etc.

Table 2.4 Verification methods proposed for determining quality of stabilized soil (Hosoya et al., 1996; Halkola, 1999; Larsson, 2005).

No.	Verification test	Method	Method description, characteristics, correlation with strength, limitation, etc.
1	Laboratory test on drilled core sample	Unconfined compression test and/or Other lab tests	Retrieval of intact core of treated soil columns and store the sample under predetermined condition until laboratory testing, commonly unconfined compression test. The verification test results can be directly compared with the design assumption. Nevertheless, most of the alternate in-situ test procedures are calibrated against q_u test results on core samples.
2	Laboratory test on wet grab sample	Unconfined compression test	Retrieve "fresh" soil-binder mixture immediately after mixing by a special probe, molding it at site and store the specimen until laboratory testing. Sampling cylinder may tend to collect unmixed cement slurry rather than soil-binder mixture.
3	Sounding	Ordinary column penetration test	A probe equipped with two opposite vanes is statically pressed down into the center of treated soil column and continuous record of resistance is taken. Commonly used for Nordic Dry Method. Applicable for $q_u < 300$ kPa down to 8 m, for $q_u < 600–700$ kPa down to 20 or 25 m if pre-bored at the center. Swedish guideline for the test is available.
4		Reverse column, penetration test	A probe attached to the wire is placed at the bottom of the treated soil column during production and left there until testing. The probe is withdrawn from the column and the continuous record of resistance can be made. Applicable for $q_u < 1200$ kPa down to 20 m. Bearing capacity formula is used to evaluate undrained shear strength of treated soil, where $N_c = 10$ (Sweden), 10–15 (Finland).
5		Standard penetration test, SPT	Driving a split sampler into soil dynamically by hammering, and measure number of blows to penetrate 30 cm. Empirical correlation between SPT N value and q_u has been reported, $q_u = \alpha N$ where $N = 25$ to 33 for soils with $q_u < 1000$ kPa.
6		Portable dynamic cone penetration test (Japan)	Driving a cone into soil by hammering, and measure blow count to penetrate 10 cm. Applicable for $q_u = 200$ to 500 kPa. Blow count Nd is correlated to unconfined compressive strength, $q_u = 29$ Nd – 258 kPa for soils with $q_u < 1000$ kPa.
7		Dynamic cone penetration test (Finland)	Driving a cone into soil by hammering and measure penetration depth for each blow. 1.12
8		Combined static-dynamic penetration test (Finland)	DCP index is correlated to CBR, $CBR = 292/DCP$ for the 60° cone angle. Combination of static penetration and hammering test. During penetration, the rod is rotated continuously by 12 rpm and torque is measured to calculate shaft friction. Applicable for $q_u < 4$ MPa.

QC/QA of deep mixing method   43

9	Cone penetration test, CPT	Cone is statically penetrated into ground and measure the penetration resistance, skin friction and pore water pressure. The undrained shear strength is corrected by $c_u = (q_t - s_{v0})/N_{kt}$, where, c_u is undrained shear strength, s_{v0} and N_{kt} are total overburden pressure and cone factor, respectively.
10	Rotary penetration sounding Test, RPT (Japan)	A sensing rod equipped with a special drilling bit is attached at the bottom end of drilling shaft is drilled into the treated soil column, and measure drilling speed R, rotation n, thrust W, torque T and water pressure u at the drilling bit. Unconfined compressive strength qu is correlated to measured data by $q_u = K \cdot R^a \cdot n^b \cdot W^c \cdot T^d$, where K, a, b, c, d are constants.
11	Automatic Swedish weight sounding test, A-SST (Japan)	A screw point connected to a series of rods is driven statically into the ground to measure the number of half-rotations for every 25-cm penetration. Applicable for $qu < 500$ kPa. The equivalent number of rotations for 1-meter penetration, N_{sw} is converted to shear strength of column.
12	Column vane test (Finland)	Diameter of the vane is 130 or 160 mm and the height is one half of the diameter. Applicable for $q_u < 400$ kPa.
13	PS logging	P- and S-wave velocities are measured either by down hole test or suspension method. Their distributions with depth reflect the uniformity of treated soil columns. Elastic modulus of the treated soil column at small strain can also be calculated from these velocities.
14	Electro-magnetic logging	Measuring electrical and magnetic properties of the ground to identify the soil layering, cavities and underground utilities. Application of these imaging techniques to the deep mixed ground seems still be in the research stage.
15	Pressure meter test	A cylindrical probe is expanded radically onto the borehole wall and measure the pressure and radial displacement. Elastic modulus and the strength of the soil are evaluated by the measurements.
16	Non-destructive tests at top of a column	Hammering top surface of column and measuring the reflected waves at the top surface to assess the continuity of treated soil columns or the shape of as-built columns. Applicable for more than 4 m long and with $q_u > 1$ MPa.
17	Impact acceleration test	A rammer is free fallen onto the treated soil ground surface and "impact acceleration" is measured, and to converted to unconfined compressive strength.
18	Plate loading test on top of a column	Rigid plate is statically loaded by step-wise to measure bearing capacity and deformation characteristics.
19	Full-scale load test of a single column	Pile load test or compression test is carried out in-situ column or extracted column to determine the load bearing capacity of single treated soil column. These tests have been conducted so far for the research purpose and not for daily QA/QC undertakings.
20	Extraction of full-scale treated soil column	Visual observation of whole column and testing. Retrieve of the full-scale treated soil column by huge sampler and test by pocket vane or by phenolsulfonphthalein to determine the uniformity.

2.3.8 Quality verification

2.3.8.1 Verification methods

During and after the construction of improved ground, the quality of stabilized soil column/element shall be verified prior to the construction of superstructure in order to confirm the design quality, such as continuity, uniformity, strength, permeability or dimension of stabilized soil column/element. There are several verification methods including the cone penetration test and the vane penetration test, as summarized in Table 2.4 (Hosoya et al., 1996; Halkola, 1999; Larsson, 2005). However, in Japan, the verification is usually carried out by means of visual inspection and testing on the boring core samples of production columns/elements.

2.3.8.2 Position of core boring

The boring core samples should be taken from each representative site and soil layer, as shown in Figure 2.20 (The Building Center of Japan and the Center for Better Living, 2018). In the case of large DM project, the improved area can be spatially classified into zones depend on the ground condition and improvement condition (e.g. mixing condition and layout of columns/elements), inspection zones A and B as an example. In each zone, the ground is usually stratified with several layers with different soil properties. The boring core samples are taken from each inspection zone and inspection soil layer so that the quality of whole improved ground can be evaluated precisely.

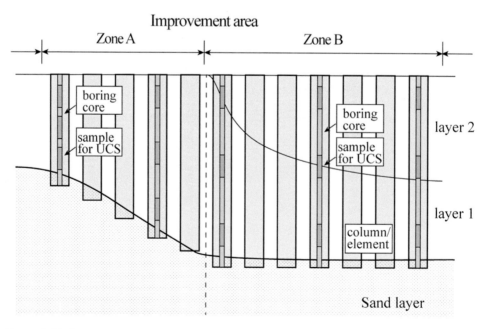

Figure 2.20 Representative ground and site (After The Building Center of Japan and the Center for Better Living, 2018).

2.3.8.3 Frequency of core boring

The frequency of boring core is determined depend on the size of project, the volume of stabilized soil, and the type and importance of superstructure. In the case of on-land works in Japan, three boring cores are generally taken in the case where the total number of columns/elements is less than 500. When the total number exceeds 500, one additional boring core boring is taken for every further 250 columns/elements. In the on-land works for residential house, one boring core boring is generally taken every 100 columns/elements (The Building Center of Japan and the Center for Better Living, 2018). In the marine work, the boring core is taken every 10,000 m^3 in stabilized soil volume for small size project and every 50,000 m^3 for large-scale project exceeding 100,000 m^3 in stabilized soil volume.

The reliability and accuracy of unconfined compressive strength on boring core sample depend upon the quality of boring core sample, and it depends upon the drilling and coring method and drillers' skill. There are many type of core boring techniques, the applicability of which are summarized in Table 3.3 in Chapter 3. The double-tube

Figure 2.21 Core boring machine.

Figure 2.22 Example of the boring core sample.

core sampler or triple-tube core sampler should be used for stabilized soil column/element due to its high strength. It is advisable to use large diameter sampler in order to take high-quality samples (Enami and Hibino, 1991).

Boring core sample are taken throughout the depth in order to verify the uniformity and continuity of the stabilized soil column/element by visual inspection irrespective of the marine and on-land works. Figures 2.21 and 2.22 show a platform and core boring machine, and an example of boring core sample of cement stabilize soil, respectively.

2.3.8.4 Quality verification of boring core sample

The quality and uniformity of the boring core sample of stabilized soil column/element is evaluated by the visual inspection and the Rock Quality Designation (RQD) index. The RQD index is defined as the borehole core recovery percentage incorporating only pieces of solid core that are longer than 100 mm in length measured along the centerline of the core, as Equation (2.4). The RQD index was originally applied for evaluating the quality of rock sample, as shown in Table 2.5. The required RQD is specified in each project, while it should be larger than 90% for sandy soil and larger than 95% for clay in Japan (The Building Center of Japan and the Center for Better Living, 2018). The RQD index has close relationship with the variation of unconfined compressive strength of boring core sample, as shown in Figure 2.23 (Kawamura et al., 2001):

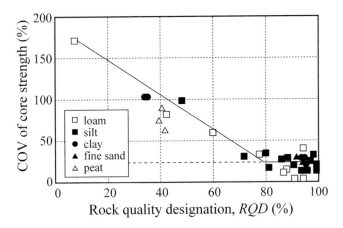

Figure 2.23 Relationship between RQD and COV of boring core strength (Kawamura et al., 2001).

Table 2.5 RQD index and rock quality

RQD	Description of rock quality
0%–25%	Very poor
25%–50%	Poor
50%–75%	Fair
75%–90%	Good
90%–100%	Excellent

$$RQD = \frac{\sum \text{length of core pieces} > 10\,\text{cm}}{\text{Total corerun length}} 100\% \qquad (2.4)$$

2.3.8.5 Quality verification by laboratory test

The engineering properties of the stabilized soil are usually evaluated based on unconfined compressive strength on boring core samples at 28-day curing. In Japan, three core barrels of 1 m length are selected from the boring core samples at three different depths, and three specimens are taken from each core barrel and subjected to an unconfined compression test (Figure 2.24). The other engineering properties such as compressibility properties and permeability are also measured on the specimen, but they can be roughly estimated by the unconfined compressive strength together with the accumulated correlation.

The testing procedure specified for ordinary cohesive soils is adopted for stabilized soil in many countries and organizations. In the test, the boring core sample is cut and trimmed to make a specimen having the same diameter as the boring core sample and the height-to-diameter ratio of 2. The sample is subjected to vertical loading with the loading speed of 1%/min. The stress and strain relationship is measured

Figure 2.24 Unconfined compression test apparatus.

to obtain the unconfined compressive strength, q_u, the axial strain at failure, m_f, the elastic modulus, E_{50}, and the residual strength, q_{ur}. It is desirable to measure the water content of the piece of stabilized soil sample after the unconfined compression test to obtain the relationship with the q_u. This relationship can provide useful and helpful inspiring in the quality assurance, especially when discussing the reason behind the non-compliant stabilized soil column/element.

2.3.8.6 Evaluation of unconfined compressive strength

The measured strength should be evaluated by an suitable manner that should be consistent with the design concept and design requirement in the geotechnical design. The measured strength is statistically evaluated to study if the stabilized soil column/element is compliant with the contract specification. In the practice, the average strength, q_{uf}, and the coefficient of variation, COV, of the strength are investigated and compared with the contract specification. The detail evaluation procedure and acceptance criteria of the field strength, q_{uf} are specified for building application in Japan (The Building Center of Japan and the Center for Better Living, 2018). The procedure is briefly explained as follows. The average unconfined compressive strength, q_{uf}, of barrel of boring core is calculated by three measured data that is a representative value for the barrel. The standard deviation, Σ_d, is calculated by the average strengths of barrels. They are evaluated by the criteria of Equation (2.5).

a. In the case where the number of boring core barrels is less than 25:

$$\overline{q_{uf}} \geq q_{ua} = q_{uck} + k_a \times \sigma_d \tag{2.5a}$$

Table 2.6 Magnitude of k_a for the probability of 10%.

No. of boring core barrels	1	2	3	4–6	7–8	9 and more
Coefficient, k_a	1.9	1.7	1.6	1.5	1.4	1.3

Table 2.7 Magnitude of k_b.

Probability (%)	5	10	20	30
Coefficient, k_b	1.65	1.3	0.8	0.5

In the case where the number of boring core barrels is larger than 25:

$$\overline{q_{uf}} \geq q_{ua} = q_{uck} + k_b \times \sigma_d \qquad (2.5b)$$

where

k_a: coefficient
k_b: coefficient
q_{ua}: acceptance strength (kN/m^2)
q_{uck}: design standard strength (kN/m^2)
q_{uf}: average unconfined compressive strength (kN/m^2)
Σ_d: standard deviation of measured unconfined compressive strength (kN/m^2)

The amount of k_a and k_b depends on the number of test data, as shown in Tables 2.6 and 2.7, for the probability of 10%.

This evaluation procedure is also applicable when the average strength and *COV* are calculated by all measured data on boring core specimens instead of those of boring core barrels.

2.3.9 Rectification of non-compliant column/element

Any stabilized soil column/element is treated as non-compliant with the contract if any construction parameters or the quality of stabilized soil column/element is not satisfied with the contract specification. In this case, review and rectification of non-compliant deep mixing work are essential. The contractor shall review the deep mixing equipment and procedure based on the recorded production log data to investigate the cause for the non-compliant column/element and modify them accordingly. The influence of the non-compliant column/element to the performance and stability of the improved ground should be evaluated to determine the necessity and remediation measurement. In the case of non-compliant DM, rectification of non-compliant DCM works is necessary. Several compensations stabilized columns/elements are usually produced close to the non-compliant column/element to reinforce it and assure its performance, when it is thought that the non-compliant column/element is caused by a mixing condition peculiar to the column/element. In the case where it is thought that the cause of the non-compliant column/element is caused by construction method itself, compensation-stabilized columns/elements should be produced at the certain area.

REFERENCES

Aoyama, K., Miyamori, T., Wakiyama, T. and Kikuchi, D. (2002) The influence of physical property on improved soil character. *Journal of the Japan Society of Civil Engineers.* No. 721/VI-57, pp. 207–219 (in Japanese).

Cement Deep Mixing Method Association (1999) *Cement Deep Mixing Method (CDM), Design and Construction Manual.* Cement Deep Mixing Method Association, 192p. (in Japanese).

Cement Deep Mixing Method Association (2006) *CDM-LODIC method Technical Manual.* Cement Deep Mixing Method Association, 50p. (in Japanese).

Coastal Development Institute of Technology (2008) *Technical Manual of Pneumatic Flow Mixing Method, revised version.* Daikousha Publishers, 188p. (in Japanese).

Coastal Development Institute of Technology (2019) *Technical Manual of Deep mixing method for marine works, revised version.* Daikousha Publishers, 315p. (in Japanese).

Dry Jet Mixing Method Association (2006) *Dry Jet Mixing (DJM) Method Technical Manual.* Dry Jet Mixing Association. (in Japanese).

EuroSoilStab (2002) Assessment of key properties of solidified fly ash with and without sodium sulfate. *Design Guide Soft Soil Stabilization. EC project BE96–3177*, 94p. (in Japanese).

Federal Highway Administration (2013) Federal Highway Administration Design Manual_ Deep Mixing for Embankment and Foundation Support.

Halkola, H. (1999) Keynote lecture: Quality control for dry mix methods. *Proceedings of the International Conference on Dry Mix Methods for Deep Stabilization.* Stockholm, pp. 285–294.

Hirano, S., Mizutani, Y., Nakamura, H., Shimomura, S. and Sasada, H. (2015) Quality improvement of deep mixing method by dispersant additives. *Proceedings of the 50th Annual Conference of the Japanese Society of Soil Mechanics and Foundation Engineering*, pp. 881–882 (in Japanese).

Hosoya, Y., Nasu, T., Hidi, Y., Ogino, T., Kohata, Y. and Makihara, Y. (1996) Japanese Geotechnical Society Technical Committee Reports: An evaluation of the strength of soils improved by DMM. *Proceedings of the 2nd International Conference on Ground Improvement Geosystems.* Vol. 2, pp. 919–924 (in Japanese).

Japan Cement Association (2012) *Soil Improvement Manual Using Cement Stabilizer* (4th edition). Japan Cement Association, 442p. (in Japanese).

Kamimura, K., Kami, C., Hara, T., Takahashi, T. and Fukuda, H. (2009) Application example of deep mixing method with reduced displacement due to mixing (CDM-LODIC). *Proceedings of the International Symposium on Deep Mixing and Admixture Stabilization*, pp. 535–540.

Kawamura, M., Hibino, S., Tamura, M., Fijii, M. and Watanabe, K. (2001) Perspective of performance-based deep-mixing method. *Journal of the Japanese Society of Soil Mechanics and Foundation Engineering, "Tsuchi to Kiso".* Vol. 49, No. 5, pp. 1–3 (in Japanese).

Kitazume, M. and Imai, H. (2021) Effect of cold joint on behavior of block type deep mixing improved ground. *Proceedings of the 46th Deep Foundation Institute Conference*, 2021 (to be published).

Kitazume, M. and Terashi, M. (2013) *The Deep Mixing Method.* CRC Press, Taylor & Francis Group, Boca Raton, FL, 410p.

Larsson, S. (2005) State of practice report - execution, monitoring and quality control. *Proceedings of the International Conference on Deep Mixing – Best Practice and Recent Advances*, Stockholm. Vol. 2, pp. 732–785.

Mizuno, S., Sudou, F., Kawamoto, K. and Endou, S. (1986) Ground displacement due to ground improvement by deep mixing method and countermeasures. *Proceedings of the 3rd Annual Symposium of the Japan Society of Civil Engineers on Experiences in Construction*, pp. 5–12.

Mizutani, Y. and Makiuchi, K. (2003) Effects of Surfactant Additive on Strength Stability of Soil Cement Pile. *Proceedings of the 48th National Symposium on Geotechnical Engineering*, pp. 45–52 (in Japanese).

Nakamura, R. 1977a. Deep mixing method with cement slurry (1) - Field test at Daikoku Pier of Yokohama port and design. *Journal of Land Reclamation and Dredging*, No. 78, pp. 32–55 (in Japanese).

Nakamura, R., 1977b. Deep mixing method with cement slurry (2) - Field test at Daikoku Pier of Yokohama port and design. *Journal of Land Reclamation and Dredging*, No. 79, pp. 23–38 (in Japanese).

Nozu, M., Anh, N.T., Shinkawa, N. and Matsushita, K. (2012) Remedy of deep soil mixing quality for montmorillonite clay deposited in the Mekong and Mississippi deltas. *ISSMGE-TC211, Brussels*. Vol. 2, pp. 443–449.

Nozu, M., Sakakibara, M. and Anh, N.T. (2015) Securing of in-situ cement mixing quality for the expansive soil with the Montmorillonite inclusion. *Proceedings of the Deep Mixing 2015*, San Francisco, June 2–5, pp. 845–852.

Public Works Research Center (2004) *Technical manual on deep mixing method for on land works*. Public Works Research Center, 334p (in Japanese).

Terashi, M. (2003) The state of practice in deep mixing method. Grouting and Ground Treatment. *Proceedings of the 3rd International Conference*, ASCE Geotechnical Special Publication, No. 120. Vol. 1, No. 120, pp. 25–49 (in Japanese).

Terashi, M. and Kitazume, M. (2009) Keynote lecture: Current practice and future perspective of QA/QC for deep-mixed ground. *Proceedings of the International Symposium on Deep Mixing and Admixture Stabilization*, pp. 61–99.

The Building Center of Japan and the Center for Better Living (2018) *Design and Quality Control Guideline of Improved Ground for Building, 2018*. The Building Center of Japan and the Center for Better Living, 708p. (in Japanese).

Yamada, Y. and Furuhashi, S. (1985) Deep mixing method using moderate heat cement and slag. *Ground and Construction*, Vol. 3, No. 1, pp. 57–66 (in Japanese).

Yoshida, S. (1996) Shear strength of improved soils at lap-joint-face. *Proceedings of the 2nd International Conference on Ground Improvement Geosystems*, pp. 461–466.

Yoshino, Y., Kishimoto, K., Sakaida, S. and Goto, K. (2002) Characteristics of heaved ground by deep mixing method. *Proceedings of the 57th Annual Conference of the Japan Society of Civil Engineers*, pp. 643–644 (in Japanese).

Chapter 3

Technical issues on QC/QA of stabilized soil

3.1 INTRODUCTION

The current QC/QA procedure was introduced in the previous chapter. However, it may not be practically possible to conduct the QC/QA introduced in the previous chapter in each construction site due to several reasons, such as site conditions, time and economical limitations, *etc*. It is necessary to implement QC/QA to assure the design criteria and performance of the improved ground within the field situation, time and economic condition. In this chapter, some matter that may affect the evaluation of QC/QA is briefly explained.

Since the magnitude and distribution of the earth pressures up to failure are still not well determined, detailed analysis such as finite element method (FEM) analysis should be conducted to achieve more reliable and precise design.

3.2 FIELD AND LABORATORY STRENGTHS

3.2.1 Prediction of strength

The strength of stabilized soil is affected by many factors, as described before. Some formulations of laboratory and field strength predictions are exemplified as Equation (3.1) (Tsuchida and Tang, 2012; Horpibulsuk et al., 2003; Tang et al., 2000; Yanagihara et al., 2000; Miyazaki et al., 2001; Yoshida et al., 1977).

$$\left.\begin{array}{l} q_u = aw + b \\ q_u = a/(W/C)^x + b \\ q_u = aC/w^x + b \\ q_u = \dfrac{10 \cdot Gs \cdot \rho_W \cdot K(c-c_0)}{v^3} \\ q_u = \dfrac{A}{B^{(WC/C)}} \\ q_u = \dfrac{K(C-C_0)}{(w \cdot Gs/100 + 1)^2} \\ q_u = k_c \cdot (c^* - c_0^*)Y^3 \end{array}\right\} \quad (3.1)$$

DOI: 10.1201/9781003223054-3

where
    c: binder factor (kg/m^3)
    c^*: weight ratio, $W_c/(W_s + W_c)$
    C: binder factor (kg/m^3)
    c_0: reference binder factor (kg/m^3)
    c_0^*: reference weight ratio, $W_c/(W_s + W_c)$
    C_0: reference binder factor (kg/m^3)
    Gs: specific gravity of soil particle
    q_u: unconfined compressive strength (kN/m^2)
    Vc: volume of binder (m^3)
    Vs: volume of soil (m^3)
    V_v: volume of void and water (m^3)
    w: water content (%)
    Wc: dry weight of binder (kg)
    W/C: water to binder ratio
    wc/c: clay water content to binder ratio
    Ws: dry weight of soil (kg)
    Y: volume ratio, $(Vs+Vc)/(Vs+Vc+Vv)$
    a, b, kc, x, A, B, K: parameter
    ρ_w: density of water (g/cm^3)

However, there is no widely applicable formula for estimating the field and laboratory strengths which incorporates all the relevant factors because the strength of field-stabilized soil depends on the deep mixing construction equipment and procedure and curing conditions.

3.2.2 Strength ratio of field to laboratory strengths, q_{uf}/q_{ul}

One of the essential factors in the process design is the strength ratio, q_{uf}/q_{ul}. There are a lot of database on the strength ratio. However, the field strength obviously depends on the deep mixing construction equipment and production process, and the laboratory strength also depends on the testing conditions, especially a molding technique that will be discussed in Section 3.4. It should be noted that the strength ratio in the database should be treated as an example and their applicability should be examined in the laboratory mix test and field-trial test.

Some of them accumulated in Japan are exemplified in Figures 3.1 and 3.2 (Public Works Research Center, 2004). The laboratory strength in the figure are the measured data on the samples produced according to the Japanese standard (The Japanese Geotechnical Society, 2009), and the field strength is that on the samples produced by Japanese deep mixing equipment and procedure. In the case of the wet method of on land works (Figures 3.1a and b), the q_{uf}/q_{ul} is almost 1/1.5 to 1 for clay and about 1/2 to 1 for sand. In the dry method (Figure 3.1c and d), the q_{uf}/q_{ul} is about 1/3 to 1.5 for clay and 1/3 to 1 for sand, as small as 1/2–1/5 for clay, but for sand relatively high field strength is obtained and the ratio larger than unity is often found.

In the case of marine works (Figure 3.2), on the other hand, the q_{uf} value is almost the same order with the laboratory strength, q_{ul}. irrespective of the binder type. The reason why the ratio of q_{uf}/q_{ul} is quite different in on land works, and marine works

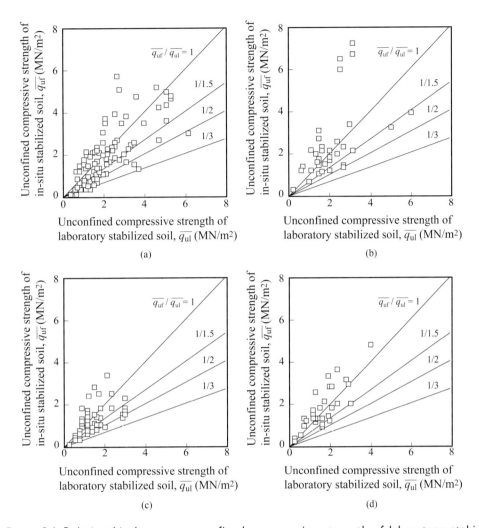

Figure 3.1 Relationship between unconfined compressive strength of laboratory stabilized soil and field-stabilized soil (Public Works Research Center, 2004). (a) Stabilized soils of on land works (wet method, clay). (b) Stabilized soils of on land works (wet method, sand). (c) stabilized soils of on land works (dry method, clay). (d) Stabilized soils of on land works (dry method, sand).

is attributed to a relatively large amount of stabilized soil and relatively good mixing degree in marine works (Kitazume and Terashi, 2013).

3.2.3 Strength deviation in field strength

Figure 3.3a shows the relationship between the average strength and coefficient of variation (COV) of field-stabilized soil using the wet method, where a stabilized soil column/element was produced by either a large-size deep mixing equipment (Cement

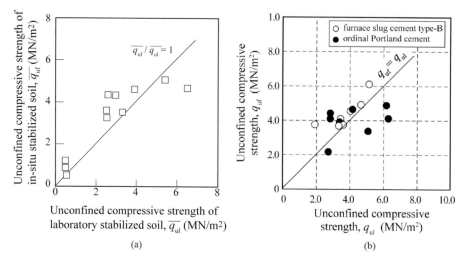

Figure 3.2 Relationship between unconfined compressive strength of laboratory stabilized soil and field-stabilized soil (wet method). (a) marine works (Noto et al., 1983). (b) marine works (Coastal Development Institute of Technology, 2019).

Deep Mixing Method Association, 1999) or small-size deep mixing equipment (The Building Center of Japan and the Center for Better Living, 2018). In the case of the small-size equipment for foundation of residential house, a large variation can be seen in the coefficient of variation (COV) when the stabilized soil strength is small, but the variation becomes small with the increase in the stabilized soil strength. In the case of large-size equipment (Cement Deep Mixing Method Association, 1999), the COV is smaller than that of the small-size equipment and also decreased with the increase in the stabilized soil strength. It can be seen that the COV is decreased to about 20% with the increase in the stabilized soil strength irrespective of the size of equipment.

Figure 3.3b shows the relationship between the COV and binder factor. Although there is a large scatter, it can be seen that the COV is gradually decreased with the increase in the binder factor, which is consistent with that in Figure 3.3a.

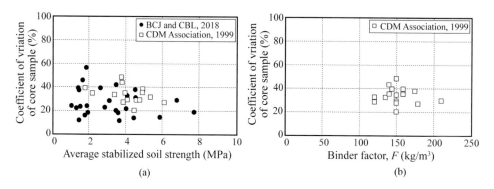

Figure 3.3 Strength and variation of field-stabilized soils. (a) Average strength and *COV*. (b) Binder factor and *COV*.

3.3 LABORATORY MIX TEST

3.3.1 Role and basic approach of laboratory mix test

The laboratory mix test is normally conducted once in a project either by the owner or by the contractor depending on the contract scheme. A design engineer uses the laboratory mix test results for assuming/establishing design parameters, and a contractor uses the same test results for planning the field-trial test or for the process design. Only when the laboratory mix test is conducted according to the standardized procedure, a certain party involved in a project can rely on the test results obtained by a different party.

However, it is found that nationwide (or regional) official standards or guidelines on the laboratory mix test are scarce. Several kinds of laboratory mix testing procedures have been adopted in each region and organizations (Terashi and Kitazume, 2009; Kitazume and Terashi, 2009; Kitazume et al., 2009a, b; The Japanese Geotechnical Society, 2009; Hirabayashi et al., 2009; Jeong et al., 2009; Kido et al., 2009, Kitazume and Nishimura, 2009; Marzano et al., 2009; Åhnberg and Holm, 2009). The prescriptions of the key elements in the laboratory mix test, such as specimen size, mixing procedures, curing conditions, and mechanical tests that follow laboratory mix tests, have been found fairly well regarded and accepted by many engineers. However, the test procedure, especially the method of preparation and curing of specimens, differs from one region to another or from one organization to the other even in the same region.

The testing procedures applied in several countries and regions can be classified into two basic concepts, as shown in Figure 3.4: (Concept A) reproduce of field mixing condition, and (Concept B) reproduce of ideal mixing condition.

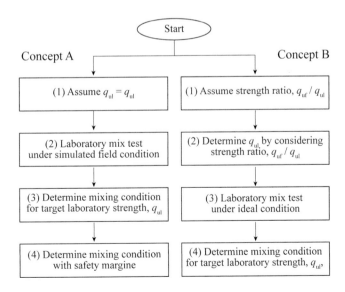

Figure 3.4 Basic concept and procedure of laboratory mix test.

Concept A, reproduce of field mixing condition.

Soil taken from construction site having a natural water content is mixed with binder having the prescribed *w/c* ratio of binder slurry. The soil and binder mixture is put into a mold in some way to achieve the expected density in the field condition. The mixture of soil and binder is usually in a plastic and cohesive state especially when the water content of original soil is relatively small and/or the amount of binder is large. The mixture with high cohesion is usually push into a mold by either static compression, dynamic compression or rodding techniques (see Section 3.4). The mixture is cured in a laboratory with/without the overburden pressure. After the unconfined compression test on the sample, the required type and amount of binder are obtained by the mixing condition that provides the q_{ul} as the same as the target q_{uf}. The amount of binder mixed in the field is usually 10%–15% more than the above by considering uncertainty in the field, which is a sort of safety margin.

The essential issue of this concept is how much this testing condition and procedure can reproduce the field mixing and curing conditions. The mixing blades and procedure in the field are obviously different from a mixer used in a laboratory. In addition, the stabilized soil in the field is cured under the overburden pressure and high temperature due to the binder hydration effect. These effects are not always reproduced in the laboratory test.

Concept B, reproduce of ideal mixing condition.

The basic concept is to product homogeneous and high reproducible stabilized soil in a laboratory. For this, original soil is usually mixed with some amount of water to become an uniform slurry with high fluidity and then mixed with binder to have uniform soil and binder mixture. The soil and binder mixture with relatively high fluidity is poured into a mold and is usually slightly compacted by the tapping method to make homogeneous and high reproducible sample. The mixture is cured in a laboratory without the overburden pressure. After the unconfined compression test on the sample, the required type and amount of binder are obtained by the mixing condition that provides the target q_{uf} estimated by the measured q_{ul} and the multiplication of the strength ratio of q_{uf}/q_{ul}. A little smaller value of the strength ratio than the measured value is usually adopted as a sort of safety margin, which requires a little more binder.

The essential issue of this concept is the strength ratio, q_{uf}/q_{ul}. The field strength, q_{uf}, is influenced by many factors and also depend on the equipment type, shape and layout of mixing blades, and mixing process. The laboratory strength, q_{ul}, also depends on the testing procedure, such as specimen size, mixing procedures, curing conditions, and the method of preparation and curing of specimens. Though there are a lot of accumulated data on the strength ratio, it should be noted that they are not always applicable in all the field conditions but can be applied to only a particular field condition.

3.3.2 Selection of soil for laboratory test and water to binder ratio of binder slurry, *w/c*

It should be noted that soil and binder are basically mixed on a horizontal plane by the rotation of vertical mixing shafts to produce a stabilized soil column/element that

Figure 3.5 Continuous mixing techniques. (a) Cutter soil mixing method (http://bauerkouhou.com/csm.html). (b) TRD method (https://www.toko-geo.co.jp/construction/show/122). (c) Power Blender method (http://www.power-blender.com).

remembers the soil stratification of original ground, and hence, the laboratory mix test should be programmed for each and all representative soil layers within the improvement depth (see Layers 1 and 2 in Figure 2.15). As the water to binder ratio of binder slurry, w/c, is ranging from 60% to 120% for many deep mixing equipment, the laboratory mix test should be programmed for the binder slurry with the w/c within the range.

However, in the case of the continuous mixing techniques, such as the cutter soil mixing method (Sakuma, 2013) (Figure 3.5a), the TRD method (Aoi et al., 1996) (Figure 3.5b) and the Power Blender method (Satou, 2009) (Figure 3.5c), soil and binder are basically mixed on a vertical plane by the rotation of horizontal mixing shafts to produce a stabilized soil panel, where stratified layers are moved vertically and mixed together. For these methods, the laboratory mix test should be programmed for the mixture of representative soil layers by taking into account the thickness of each layer. As the water to binder ratio of binder slurry, w/c, is ranging from 100% to 250% for many deep mixing equipment, the laboratory mix test should be programmed for the binder slurry with the w/c within the range.

3.3.3 Effect of specimen size

As the unconfined compression test procedure for stabilized soil is not specified, the test procedure for an ordinary cohesive soils is adopted for stabilized soil in many countries and organizations. The diameter-to-height ratio of an unconfined compression test specimen is usually specified 1:2, but the dimensions themselves are not specified in many standards. However, the diameter and height of an unconfined compression test specimen for the laboratory mix test are usually adopted 5 and 10 cm, but those for field sample are 8.6 and 17.2 cm or 11.6 and 23.2 cm depending on the size of boring core sample in Japan. The effect of specimen size on the unconfined compressive strength have been investigated, some examples will be introduced in the next section.

3.3.3.1 Strength

Figure 3.6 shows the effect of specimen size on the unconfined compressive strength on laboratory mixed Kaolin clay (w_l of 50.6%, Ip of 19.6) (Omine et al., 1998, 2005), where the clay was stabilized with ordinary Portland cement with the w/c ratio of cement slurry of 100%. The figure clearly shows that the unconfined compressive strength is decreased with the increase in the specimen size irrespective of the cement dosage. It is also found that the scatter in strength is also influenced by the specimen size, while the large-size specimen shows small variation.

A similar phenomenon is observed for the other types of soil, Naruo marine clay (w_l of 72.9%, w_p of 24.7% and Ip of 48.2) stabilized with ordinary Portland cement (Yamamoto and Miyake, 1982), and Kaolin clay (w_l of 50.6%, w_p of 19.6% and Ip of 31) stabilized with ordinary Portland cement (Hayashi et al., 1996). Figure 3.7a shows the summary of the test data, in which the strength ratio of arbitrary size specimen to that of 4 or 5 cm in diameter is plotted on the vertical axis for comparing the test data directly. The strength ratio is decreased with the increase in the specimen size irrespective of the type of soil and mixing conditions. In the figure, the specimen size effect on the Brazilian tensile strength is also plotted together. The tensile strength also shows the similar strength decrease tendency to the unconfined compressive strength.

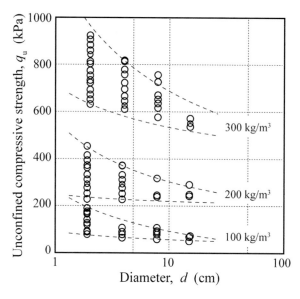

Figure 3.6 Effect of specimen size on unconfined compressive strength on laboratory stabilized Kaolin clay (After Omine et al., 2005).

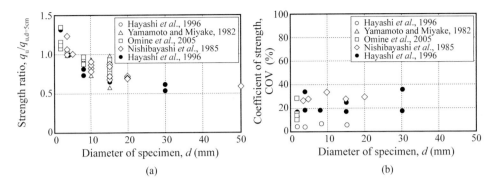

Figure 3.7 Effect of specimen size on unconfined compressive strength and coefficient of variation. (a) Strength ratio. (b) Coefficient of variation.

Figure 3.7b shows the effect of specimen size on the coefficient of variation, COV of unconfined compressive strength and Brazilian tensile strength. There is a large scatter ranging from 5% to 40% depending on the soil type and mixing conditions, and the size effect on the magnitude of the COV is almost constant irrespective of the specimen size.

Figure 3.8 shows the effect of mixing degree on the specimen size effect (Nishibayashi et al., 1985a), in which Kawasaki marine clay layer (w_l of 68.0% and w_p of 30.3%) prepared in a specimen box was stabilized by a small-size mixing device (diameter of mixing blade of 50 cm). The mixing condition was constant w/c of 100% and cement content of 20% throughout the test series, but the mixing degree is changed intentionally, the lowest mixing degree in Case 1 and the highest mixing degree in Case 3.

Figure 3.8a shows the strength ratio of the average strength of arbitrary size sample to that of 50 mm size specimen. In the figure, the strength of stabilized soil mixed by the electric mixer is also plotted, which is expected to the highest mixing degree. The figure clearly shows the strength ratio is decreased with the sample size irrespective of the mixing degree. Figure 3.8b shows the coefficient of variation with the sample size.

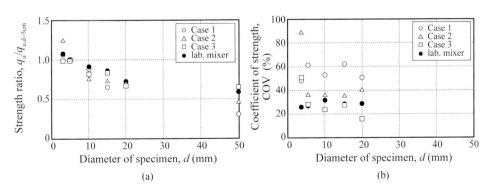

Figure 3.8 Effect of specimen size on unconfined compressive strength and Coefficient of variation (Nishibayashi et al., 1985a). (a) Strength ratio. (b) Coefficient of variation.

The figure shows a general tendency where the *COV* depends on the test condition, large *COV* in the lowest mixing degree (Case 1) and small *COV* in the highest mixing degree (Case 3), and the *COV* is decreased with the increase in the size of specimen irrespective of the mixing degree. The figures show the general phenomenon that the small-size sample provides larger strength but provide larger scatter.

3.3.3.2 Young's modulus

Figure 3.9 shows an example of the effect of specimen size on the relationship between the elastic modulus, E_{50}, and the unconfined compressive strength, q_u, (Yamamoto and Miyake, 1982). In the tests, the marine clay excavated Naruo beach (w_l of 72.9%, w_p of 24.7% and I_p of 48.2) was stabilized with ordinary Portland cement after preparing its water content of 100%. The relationship between the E_{50} and q_u is also influenced by the specimen size, while the E_{50}/q_u is increased with the increase in the specimen size. However, this tendency is no always common and depends on the soil type and laboratory and field-stabilized soils (Itou and Horiuchi, 1980).

3.3.4 Effect of molding technique

It is found that nationwide (or regional) official standards or guidelines are scarce. Several kinds of testing procedures have been adopted in each region and organizations (Terashi and Kitazume, 2009; Kitazume and Terashi, 2009; Kitazume et al., 2009a, b; The Japanese Geotechnical Society, 2009; Hirabayashi et al., 2009; Jeong et al., 2009;

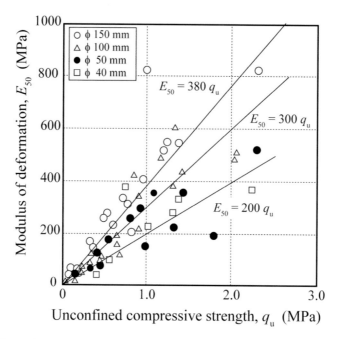

Figure 3.9 Effect of specimen size on E_{50}/q_u (Yamamoto and Miyake, 1982).

Figure 3.10 Molding techniques. (a) Tapping. (b) Rodding. (c) Dynamic compaction. (d) Static compaction.

Kido et al., 2009; Kitazume and Nishimura, 2009; Marzano et al., 2009; Åhnberg and Holm, 2009). According to the survey, several molding procedures have been adopted worldwide (Figure 3.10):

a. Tapping: Soil and binder mixture is poured into a mold, diving it into approximately three layers. Then, the mold with each soil and binder mixture layer tapped on a laboratory floor or table. The number of tapping is determined based on the local experience.
b. Rodding: Soil and binder mixture is poured into a mold, diving it into approximately three layers. Then, the each soil and binder mixture layer is poked with a rod for each layer. The number of poking and the rod diameter are determined based on the local experience.
c. Dynamic compaction: Soil and binder mixture is poured into a mold, diving it into approximately three layers. Then, each soil and binder mixture layer compacted dynamically by falling weight on the mixture layer. The weight and falling height, number of blows are determined based on the local experience.

d. Static compaction: Soil and binder mixture is poured into a mold, diving it into approximately three layers. Then, each soil and binder mixture layer is compacted statically by putting a weight on the mixture layer. The magnitude of pressure and time for press are determined based on the local experience.
e. No compaction: Soil and binder mixture is poured into a mold.

The physical and mechanical properties of stabilized soil are influenced by the molding technique (Yamamoto and Miyake, 1982; Kitazume and Terashi, 2009; Kitazume et al., 2009a, b; The Japanese Geotechnical Society, 2009; Hirabayashi et al., 2009; Jeong et al., 2009; Kido et al., 2009, Kitazume and Nishimura, 2009; Marzano et al., 2009; Åhnberg and Holm, 2009). Figure 3.11 shows an example of the effect of molding technique on the density and strength of stabilized soil for various soil types and binder types (Kitazume et al., 2015). In the figures, the undrained shear strength of fresh soil and binder mixture as an index of fluidity is plotted on the horizontal axis. On the vertical axis, the density and strength are normalized by those of sample molded by the tapping technique. There are several parameters to evaluate the applicability and reliability of the molding technique. Among them, Grisolia et al. (2013) proposed the "applicability index" for evaluating the applicability of a molding technique, which is related to "densest specimens with the highest strength" and "results repetitiveness". Kitazume et al. (2015) proposed the undrained shear strength of fresh soil and binder

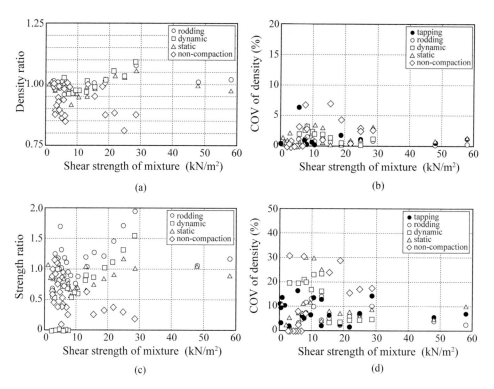

Figure 3.11 Influence of molding technique on properties of stabilized soil specimen (Kitazume et al., 2015). (a) Density ratio. (b) COV of density. (c) Strength ratio. (d) COV of strength.

mixture, and discussed the applicability and reliability of the molding technique from the view point of achieving the homogeneous and high productivity of the stabilized soil. (Figure 3.11):

1. In the case of the undrained shear strength of mixture lower or equal to 20 kN/m^2
   The rodding technique shows the highest applicability from the view point of the wet density and the unconfined compressive strength.
2. In the case of the undrained shear strength of mixture ranging from 20 to 30 kN/m^2
   The rodding and the dynamic compaction techniques show the highest applicability from the view point of density and strength. The no-compaction technique, on the other hand, shows quite small density and strength with large COV value, which means the low applicability of the technique.
3. In the case of the undrained shear strength of mixture larger than 30 kN/m^2
   The rodding technique shows better molding result than the tapping and static compaction in these cases.

The stabilized soil strength is considerably influenced by the molding technique, especially for high fluid soil and binder mixture and small shear strength of mixture. In the case where the soil and binder mixture exceeds shear strength of the order to about 50 kN/m^2, the effect of molding technique becomes very small.

As the strength ratio, q_{uf}/q_{ul}, is one of the essential parameters in the process design, their test results emphasize the importance of selecting the appropriate laboratory mixing test condition for the process design.

In principle, any molding technique can be applicable in the laboratory mix test in the Concept B, reproduce of ideal mixing condition, as far as the strength ratio, q_{uf}/q_{ul}, for the certain molding technique is obtained with certain accuracy. In Japan, the Concept B together with the tapping technique are often adopted in the deep mixing project, where a lot of database of the strength ratio have been accumulated.

Any laboratory mix test procedure can be adopted when a strength ratio is originally determined for a particular site and project. However, if using the accumulated database on the strength ratio, the laboratory mix test procedure should be the same in the database.

3.3.5 Effect of overburden pressure during curing

Field-stabilized soil produced by the deep mixing method are subjected to an overburden pressure due to the weight of soil during curing period, while that in the laboratory mix test is cured without any vertical pressure except for peat soil in Sweden.

Figure 3.12 shows the effect of the overburden pressure during the curing on the strength of the cement stabilized soil, where the Ube clay (w_l of 45.4%, w_p of 20.1% and Fc of 61.0%) was stabilized with either ordinary Portland cement or special cement (SiO_2 of 15%–20%, Al_2O_4 of more than 4.5%, CaO of 40%–70%, SO_4 of more than 4.0%) (Yamamoto et al., 2002). Figure 3.12 shows the relationship between the unconfined compressive strength at 7 days curing with the overburden pressure, σ_v' (Yamamoto et al., 2002). The figure clearly shows that the strength increases almost linearly with the overburden pressure irrespective of the type and amount of binder,

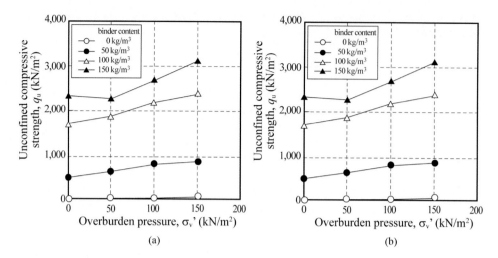

Figure 3.12 Relationship between unconfined compressive strength, q_u and overburden pressure, σ_v' (After Yamamoto et al., 2002). (a) Ordinary Portland cement. (b) Cement-based special binder.

which is due to the drainage of free water containing in the soil and binder mixture enhanced by the overburden pressure.

Figure 3.13 shows a similar relationship on the stabilized sandy soil, where sandy soil excavated Yamaguchi (ρ_s of 2.693, D_{max} of 4.8 mm, w_l of 46.6%, w_p of 22.4%, w_n of 16.4% and Fc of 18.9%) was mixed with water to adjust to prescribed water content and stabilized with ordinary Portland cement of the cement factor of 50 kg/m³ (Yamamoto et al., 2002). It can be seen that the strength at 7 days curing is increased almost linearly with the increase in the overburden pressure, as similar to those on the

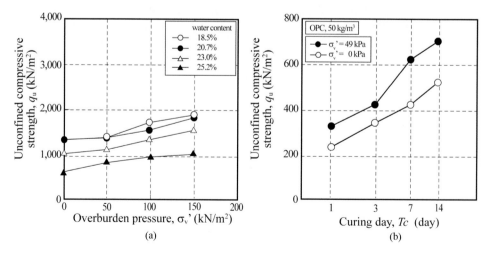

Figure 3.13 Relationship between unconfined compressive strength, q_u and overburden pressure, σ_v' (After Yamamoto et al., 2002). (a) Effect of overburden pressure. (b) Effect of curing period.

stabilized clay (Figure 3.12). Figure 3.13b shows the effect of overburden pressure and curing (loading) period on the strength. The strength is increased almost linearly with the logarithm of curing period irrespective of the magnitude of overburden pressure.

The effect of various loading patterns of overburden pressure, loading time, loading period, stepwise loading, *etc.* were discussed in detail by Yamamoto et al. (2002) and Suzuki et al. (2005).

In the laboratory mix test, any overburden pressure is not loaded on the specimen except a case for organic soil in Sweden. In Sweden, the stabilized organic soil is cured with the overburden pressure of 100 kN/m² to simulate the field curing condition. It is desirable to apply the overburden pressure during the curing to simulate the field curing condition more precisely especially in the Concept A, reproduce of field condition. However, it is practically hard to prepare loading apparatus in a laboratory. It is not necessary to adopt this effect in the Concept B, reproduce of ideal condition because this effect is already taken into account in the measured field strength, q_{uf}, and the strength ratio, q_{uf}/q_{ul}.

3.3.6 Effect of curing temperature

3.3.6.1 Temperature in ground

It is well known that the temperature in a ground is almost constant regardless of the season. However, once the deep mixing work is carried out, the stabilized soil generates heat by the hydration reaction of binder and the ground temperature is increased. Figure 3.14 shows an example of the temperatures in the ground and the stabilized soil column measured in the field-trial test, where Hamaoka sand, a fine sand (ρ_s of 2.628, w_n of 5.5%), was stabilized with special cement of the cement factor of 250 kg/m³ to produce a stabilized soil column with 1.0 m in diameter and 2.4 m in length (Enami et al., 1985a). As shown in the left figure, the temperature of the original ground at shallow depth down to about −1 m is changed seasonally. However, the temperature

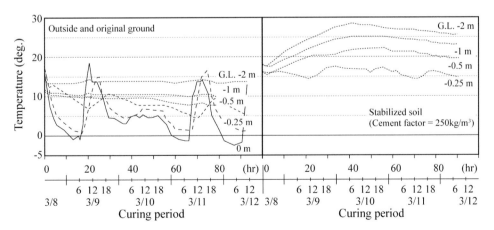

Figure 3.14 Measure temperature in ground and stabilized soil column (Enami et al., 1985a).

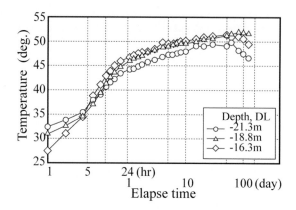

Figure 3.15 Stabilized soil temperatures (Omura et al., 1981).

of the original ground under −1 m depth is almost constant even if the outside temperature changes periodically. The right figure shows the temperature in the stabilized soil column. The figure shows the temperature is increased gradually to about 28° in Celsius at G.L. −2 m and gradually decreased due to the completion of the binder hydration.

Figure 3.15 shows another example of the ground temperature in the stabilized soil, where the block type improved ground was constructed at Yokohama Daikoku port (Omura et al., 1981). The figure shows that the temperature in the stabilized soil mass was increased to about 50°C at 40–60 days after the production. By comparing Figure 3.14, the temperature increase becomes more dominant when the stabilized soil volume becomes large.

3.3.6.2 Effects of curing temperature and period

Figure 3.16 shows the effect of curing temperature on the strength of stabilized soil, where peat (w_n of 356.9% and organic matter content of 36.0%), humus soil (w_n of 211.8% and organic matter content of 30.6%), fine sand (w_n of 5.5% and organic matter content of 1.8%) and loam soil (w_n of 109.9% and organic matter content of 17.5%) were stabilized with special cement in a laboratory (Enami et al., 1985a). It can be seen that the high temperature brings the high strength of the peat, humus soil and fine sand, especially the peat and fine sand. In the case of loam, the effect of curing temperature on the strength is quite small.

Figure 3.17 shows another example of the effect of temperature on the strength of stabilized soil measured in a laboratory test, in which Kawasaki clay with an initial water content of 55.9% was stabilized with cement of the cement factor of 160 kg/m^3 and w/c ratio of 60% and cured either in a wet box or underwater at several temperatures (Omura et al., 1981). The figure clearly shows the unconfined compressive strength is increased almost laniary with the logarithm of curing period irrespective of the curing temperature. The unconfined compressive strength of sample cured at 50°C for 60 days shows about 30% larger than that cured at 20°C.

Technical issues on QC/QA of stabilized soil 69

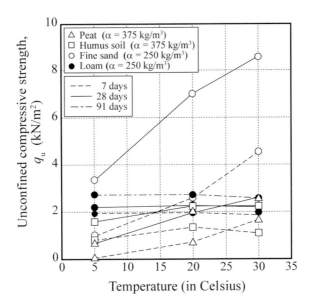

Figure 3.16 Effect of curing temperature on strength of stabilized soil (Enami et al., 1985a).

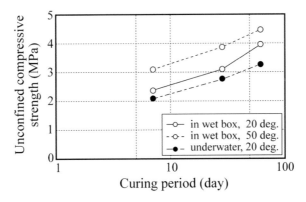

Figure 3.17 Effect of curing temperature on strength of stabilized soil (Omura et al., 1981).

Figure 3.18 shows the strength increase of stabilized soil with time, where the Kawasaki clay (w_p of 24.0% and w_l of 53.1%) was stabilized with ordinary Portland cement of several binder contents and cured at several temperatures (Kitazume and Nishimura, 2009). The figure shows the higher curing temperature and longer curing period giving larger strength.

Figures 3.18 are replotted to show the relationship between the curing temperature and strength in Figure 3.19 (Kitazume and Nishimura, 2009). The figure shows that strength is increased almost linearly with the increase in the curing temperature, and the strength increase phenomenon is more dominant for large amounts of binder.

70 QC/QA of the Deep Mixing Method

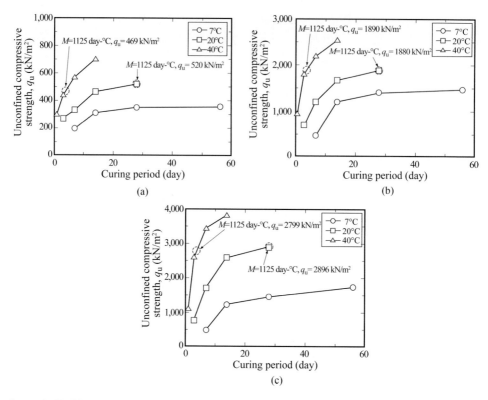

Figure 3.18 Effect of curing temperature and period on strength of stabilized soil (Kitazume and Nishimura, 2009). (a) Binder content of 5%. (b) Binder content of 10%. (c) Binder content of 15%.

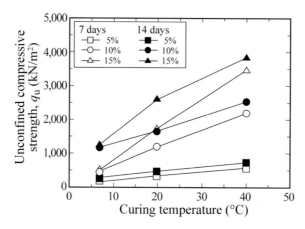

Figure 3.19 Effect of curing temperature on strength of stabilized soil (Kitazume and Nishimura, 2009).

It has been well established that higher curing temperature and longer curing period lead to larger strength of stabilized soils. The Japanese Geotechnical Society Standards stipulate 28-day curing at 20°C±3°C as the default condition. In colder or tropical regions, however, local environment may make such a 'standard condition' less relevant. Databases of strength–temperature and strength–time relationships then serve as a guide to design rational a laboratory mix test program.

It is desirable to cure the stabilized soil specimen under the expected field temperature, as exemplified in Figures 3.16 and 3.18 to simulate the field condition more precisely, especially in the Concept A, reproduce of field condition. However, it is practically hard to prepare temperature control chamber in a laboratory from the view point of economics. It is not necessary to adopt this effect in the Concept, reproduce of ideal condition, because this effect is also already taken into account in the measured field strength, q_{uf}, and the strength ratio, q_{uf}/q_{ul}.

3.3.6.3 Maturity

There are several proposals on the maturity, M, as shown in Equations (3.2) (Kitazume et al., 2009b; Babasaki et al., 1996; Nakama et al., 2003; Åhnberg and Holm 1984), which incorporates the effects of curing temperature and curing period.

$$M = \sum (T - T_0) \times t_c \tag{3.2a}$$

$$M = 2.1^{\frac{(T-T_0)}{10}} \times t_c \tag{3.2b}$$

$$M = \{20 + 0.5 \times (T - 20)\}^2 \times \sqrt{t_c} \tag{3.2c}$$

$$M = \int_0^{t_c} 2 \times \exp\left(\frac{T - T_0}{10}\right) dt \tag{3.2d}$$

where
  M: maturity (°C - day)
  t: day
  t_c: curing period (days)
  T: curing temperature (°C)
  T_0: reference temperature (−10°C)

Figure 3.19 is replotted to show the relationship between the Maturity and the stabilized soil strength, where the Maturity is formulated as Equation (3.2) in Figure 3.20. The figure clearly shows an unique linear relationship between the Maturity and the strength for each binder content irrespective of the curing temperature and period.

Figure 3.21 shows a similar relationship for various soil types, in which three types of soil were stabilized and cured at 5°C, 20°C and 30°C in laboratory mix tests (Enami et al., 1985a). The figure shows the unconfined compression test, q_u, has a linear relationship with the logarithm of Maturity, M (Equation 3.2d), irrespective of soil type

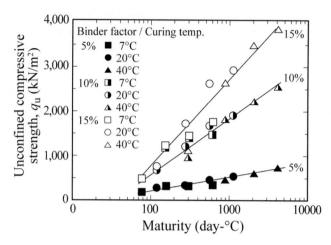

Figure 3.20 Relationship between unconfined compressive strength and Maturity (Kitazume and Nishimura, 2009).

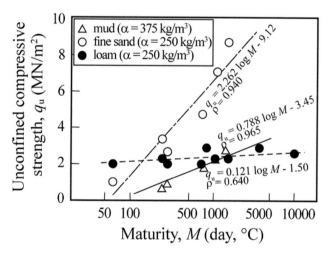

Figure 3.21 Relationship between unconfined compressive strength and maturity (Enami et al., 1985a).

and amount of binder. In the case of the mud and fine sand, the strength is increased rapidly with the M but slightly increased with the increase in the M in the case of the loam.

As shown in Figures 3.20 and 3.21, the Maturity brings the q_u data points broadly along unique lines, each of which represents different type of soil type and type and amount of binder. One of the potential applications is to estimate the strength at the 28th day, 20°C from the shorter-term tests at higher curing temperature. Equation (3.2d) implies that the 3.8 days at 40°C is equivalent to 28 days at 20°C in terms of the Maturity.

3.4 SELECTION OF DEEP MIXING EQUIPMENT

3.4.1 Factors influencing mixing degree

The deep mixing is carried out by a locally available mixing equipment from many variations, as shown in Figures 1.4–1.8. There is a tremendous difference in the level of sophistication in the mixing equipment, as already shown in Chapter 2. According to the previous research studies and field experiences, the average strength of stabilized soil column/element is increased and the coefficient of variation is decreased with the increase in the mixing degree irrespective of soil type, type of deep mixing construction equipment and mixing condition. The mixing degree is influenced by many factors, such as the shape and layout of the mixing blades, position of binder injection nozzles and also soil type and amount of binder (Nakamura and Matsushita, 1982; Nakamura et al., 1982).

A lot of research studies were carried out to investigate several aspects in regard to operation techniques on the quality of stabilized soil in Japan (e.g. Abe et al., 1997; Enami et al., 1985a, 1986b; Mizuno et al., 1986; Nishibayashi et al., 1985b; Kusakabe et al., 1996; Saito et al., 1981a, 1981b). They revealed that the mixing degree is influenced by not only the shape and layout of mixing blade but also the mixing procedure (penetration and withdrawal speeds of mixing blades, rotation speed of mixing blade, binder injection pressure, *etc*.). A deep mixing construction equipment consisting of suitable mixing tool and mixing plant should be selected for each construction condition and its applicability and efficiency should be investigated in field-trial test. Some of the key factors influencing the mixing degree will be introduced in the following sections.

3.4.1.1 Influence of number of mixing shafts

The influence of number of mixing shafts on the strength of stabilized soil was investigated in the laboratory model test (Nishibayashi et al., 1985b). The clay excavated from the Tokyo bay (w_p of 20.9% and w_l of 38.7%) was stabilized with ordinary Portland cement of the cement factor of 160 kg/m^3 and w/c of 80% by the model deep mixing equipment with 0.5 m diameter of mixing blade. Figure 3.22 shows a comparison of the test results by the single mixing shaft equipment and the four mixing shafts equipment. It is found the strength of stabilized soil by the four mixing shafts is larger than those by the single mixing shaft. It might be due to the phenomenon where soils are disturbed and moved by the mixing blades rotating to opposite direction and are collided at the overlap portion of the mixing blades to increase the mixing degree.

3.4.1.2 Influence of type and shape of mixing blade

Figure 3.23 shows the strength distribution along the depth, in which the ground was stabilized with special cement of the cement factor of 250 kg/m^3 by two types of mixing blade; horizontal-type (ordinal type) mixing blade and open-type mixing blade(Abe et al., 1997). Although there is a large scatter in the strength, the field strength obtained

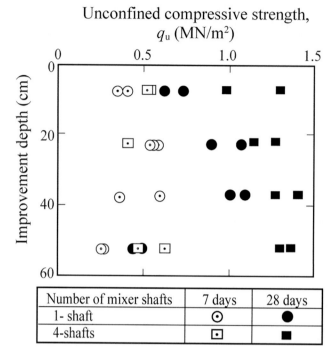

Figure 3.22 Comparison of strength of stabilized soil by single mixing shaft and four mixing shafts equipment (Nishibayashi et al., 1985b).

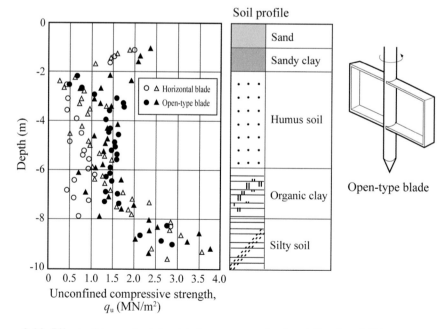

Figure 3.23 Effect of type of mixing blade on strength of stabilized soil (Abe et al., 1997).

by the open-type mixing blade is larger than those by the normal horizontal-type mixing blade. This difference is dominant in the Humuc soil layer.

Figure 3.24 shows the frequency of stabilized soil strength of Humic soil at the depth −2 to −6 m at 28 days curing. The average strength of the horizontal-type mixing blade is 0.79 MN/m², but one of the open-type mixing blade is 1.34 MN/m² which is about 70% larger than the horizontal-type mixing blade. The COV of the open-type mixing blade, 24%, is smaller than the horizontal-type, 37%.

Figure 3.25 compares the effect of the free blade on the strength, in which loam (w_n of 105.5%) was stabilized with special cement of the cement factor of 250 kg/m³ by model deep mixing equipment having two types of mixing tool, (A) and (B) (Enami et al., 1985b). In the case of the mixing tool (B), there is a free blade installed close to

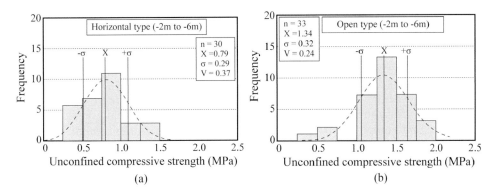

Figure 3.24 Effect of mixing blade on frequency of strength of stabilized soil (Abe et al., 1997). (a) Horizontal-type mixing blade. (b) Open-type mixing blade.

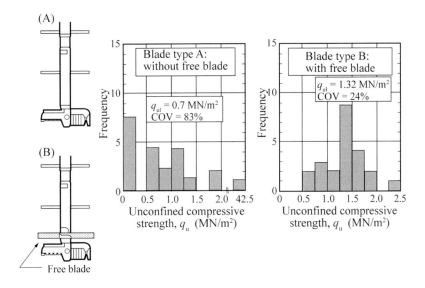

Figure 3.25 Comparison of strength of stabilized soil by mixing equipment with and without free blade (Enami et al., 1985b).

the bottom excavation blade. The free blade is a little longer than the mixing blade, which stays without rotating in the ground and to prevent the entrained mixing phenomenon. The figure shows that the strength of stabilized soil by the mixing tool without the free blade (A) is relatively small while those by the mixing tool with the free blade (B) is quite large. It clearly reveals the efficiency of the free blade to prevent the entrained mixing phenomenon and increase the strength of stabilized soil.

3.4.1.3 Influence of diameter of mixing blade

Many deep mixing equipment install mixing blade with 1.0–1.3 m in diameter. It is obvious that deep mixing equipment with large mixing blade diameter can produce a large-size stabilized soil column/element by a single stroke, which is economical. However, the required torque driving the mixing blade is increased with the increase in the diameter of mixing blade. It becomes difficult to spread the binder slurry injected from the nozzle on the mixing shaft spatially and uniformly in the cross section of the column/element when the mixing blade diameter becomes large, which causes the strength variation large in the cross section of the column/element.

Figure 3.26 shows an example of the effect of mixing blade diameter on the strength distribution within the column (Isobe et al., 1996). A sandy ground was stabilized by a deep mixing equipment with a single mixing shaft, where the cement slurry was injected during the withdrawal stage (withdrawal injection method). After the production of stabilized soil columns, the ground was excavated to expose the top of the stabilized soil columns and boring core samples were taken at the column center, 1/4 diameter and 3/8 diameter from the column center to investigate the strength distribution along the depth and in the cross section of the column.

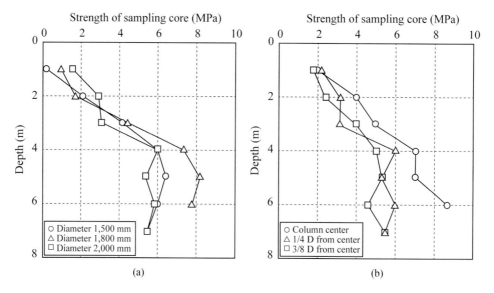

Figure 3.26 Influence of mixing blade diameter on strength of stabilized soil (Isobe et al., 1996). (a) Strength distribution along depth (at 1/4 D from column center). (b) Spatial strength distribution (by 2.0 m diameter mixing blade).

Figure 3.26a shows the effect of the mixing blade diameter on the strength distribution along the depth at the 1/4 diameter from the column center. The strength distribution along the depth is mainly caused by the difference in the soil type and condition. Though the strength is not always consistent with the mixing blade diameter, it can be seen that the strength is decreased with the increase in the diameter of mixing blade. Figure 3.26b shows the spatial distribution of the strength in the column that was produced by the mixing equipment with mixing blades of 2.0 m in diameter. It can be seen that the strength at the column center is the largest and those at the 1/4 and 1/8 diameter from the column center are smaller than that at the column center, which might be caused by the less cement and/or less mixing at the periphery of the column.

According the test results, it should be noted that the effect of the mixing blade diameter on the strength distribution within the stabilized soil column/element should be considered to select suitable deep mixing equipment.

3.4.1.4 Influence of penetration speed of mixing tool

Figure 3.27 shows an example of the influence of the penetration speed of mixing tool on the strength of stabilized soil (Enami et al., 1985b). In the field test, the Konosu clay (w_p of 50.1% and w_l of 149.2%) and Funabashi loam (w_p of 81.3% and w_l of 146.3%) were stabilized with special cement of the cement factor of 300 kg/m³ by a small-size deep mixing equipment, where the penetration speed of mixing tool was changed from 0.5 to 1.0 m/min. while the withdrawal speed and rotation speed of mixing tool were kept constant at 1.0 m/min. and 60 rpm, respectively. The figure shows that the average unconfined compressive strength is decreased very rapidly with the increase in the penetration speed irrespective of the type of soil. The figure also shows that the coefficient of deviation of the column strength, figures in the parentheses, becomes large as the penetration speed is increased.

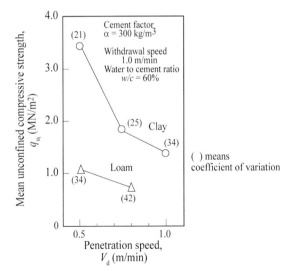

Figure 3.27 Influence of penetration speed of mixing tool on strength of stabilized soil (Enami et al., 1986b).

3.4.2 Required blade rotation number

In Japan, an index named blade rotation number, T has been introduced to evaluate (quantify) the mixing degree that is defined as Equation (2.3) in Chapter 2. There are many case studies on the relationship between the blade rotation number, T and the strength and strength variation of stabilized soil column.

3.4.2.1 Influence of blade rotation number in laboratory model tests

Figure 3.28 shows an example of the influence of the blade rotation number to the strength of stabilized soil measured in a laboratory model test (Nishibayashi et al., 1985b). The clay excavated from the Tokyo bay (w_p of 20.9% and w_l of 38.7%) was stabilized with ordinary Portland cement of the cement factor of 160 kg/m³ and w/c of 80% by a model deep mixing equipment with mixing blade of 0.5 m in diameter. As the binder was injected in the penetration stage, the blade rotation number, T, was 107 for the test of the low rotation speed and 140 for that of the high oration speed, and the penetration and withdrawal speed of the mixing tool were 0.5 a 1.0 m/min, respectively. The figure clearly shows that the strength of the stabilized soil of the large blade rotation number is larger than those of the small number. This difference still remained, even in the relatively long curing period.

Fujii et al. (2004) performed the large-scale laboratory model tests where two types of soil ground, sand (sand content of 89%, silt of 6% and clay of 4%) and loam (w_l of 122.1%, w_p of 63.1% and Ignition loss of 13.8%) grounds, were prepared. A small-size deep mixing equipment was used to produce the stabilized soil columns with 0.6 m in

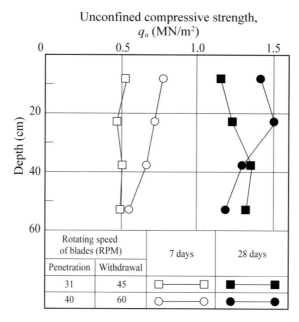

Figure 3.28 Influence of blade rotation number to strength of stabilized soil (Nishibayashi et al., 1985b).

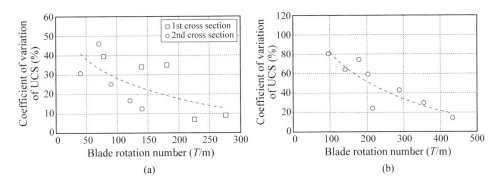

Figure 3.29 Relationship between coefficient of variation and blade rotation number. (After Fujii et al., 2004). (a) Sandy ground. (b) Loam ground.

diameter and 1.5 m in length in grounds. In the tests, the blade rotation number was changed from 38 to 276 for the sandy ground and from 108 to 540 for the loam ground. After the production of columns, boring core samples were taken from the grounds at about −0.5 and −1.0 m depth for the sand ground and −0.5 m depth for the loam ground for the unconfined compression test. Figure 3.29 shows that the coefficient of variation, COV, of the strength of the column is decreased with the increase in the blade rotation number irrespective of the soil type. It is also found that the blade rotation number of about 150 for the sand and 400 for the loam is necessary to achieve the COV lower than 25%.

3.4.2.2 Influence of blade rotation number in field test

Figure 3.30 shows similar relationship between the blade rotation number and the average strength and strength variation obtained in the five field tests (Yoshizu, 2014). In the tests, several types of soil, clay (case 401), Kanto loams (cases 402 and 403) and sandy silts (cases 404 and 405), were stabilized with special cement by a deep mixing

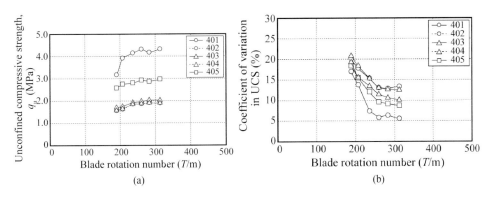

Figure 3.30 Relationship between blade rotation number and unconfined compressive strength and coefficient of variation (Yoshizu, 2014). (a) Strength. (b) Coefficient of variation.

equipment with the mixing blade of 0.6 m in diameter. The binder factor was changed 300 and 350 kg/m^3 for each soil type, while the water to binder ratio of binder slurry, *w/c*, was also changed 60% and 65%. After the production of the column, boring core samples were taken at a specific depth and subjected to the unconfined compression test. Figure 3.30a clearly shows that the average strength is increased with the increase in the blade rotation number but becomes almost constant when the blade rotation number exceeds about 250, irrespective of the soil type and mixing condition. Figure 3.30b shows the relationship with the coefficient of variation in the strength, where the *COV* is decreased with the increase in the blade rotation number but becomes almost constant when the number exceeds about 250.

3.4.2.3 Influence of blade rotation number in field actual works

Figure 3.31 shows the relationship between the blade rotation number and coefficient of variation of the strength of the field-stabilized soil strength, which were accumulated data in the field actual works for various types of deep mixing equipment and mixing process (Nakamura et al., 1982). In the figure, the laboratory mix test result is also plotted by estimating the *T* of 4,000. The magnitude and variation of *COV* are quite large when the blade rotation number, *T* is small, and are decreased almost linearly with the logarithm of the blade rotation number.

Figures 3.28–3.31 show the clear tendency in the laboratory model test, field test and field actual works where the average strength of stabilized soil column is increased, and the coefficient of variation is decreased with the increase in the blade rotation number irrespective of soil type, mixing equipment and mixing condition. However, the quantitative evaluation is different depending on them (Nakamura *et al.*, 1982). According to the accumulated research studies and investigations, the minimum blade rotation number is specified around 270 and 350 for Japanese wet and dry methods, CDM method and DJM method, respectively, to assure sufficient mixing degree (Cement Deep Mixing Method Association, 1999; Coastal Development Institute of Technology, 2019; Public Works Research Center, 2004).

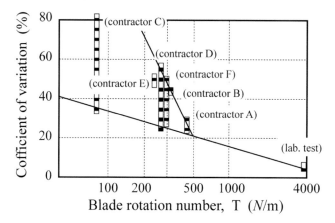

Figure 3.31 Relationship between blade rotation number and *COV* of strength of field-stabilized soil (Nakamura et al., 1982).

It should be noted again that it is obvious that the required blade rotation number depends on the specification of deep mixing equipment and production process. Contractors should accumulate the field data to obtain the required blade rotation number for each deep mixing equipment.

3.4.3 Stabilization at shallow depth and influence of ground heaving

3.4.3.1 Basic production procedure and effect of sand mat

It should be noted that some amount of injected binder slurry flow out on the ground surface along the mixing shaft without mixing with in-situ soil when the amount of binder slurry large. This phenomenon becomes dominant in the stabilization at the shallow depth due to the small overburden pressure, which causes strength decrease there rather than that at the deeper depth. The binder slurry flowed out on the ground may cause water pollution of surrounding in the case of marine works. Therefore, binder injection is terminated at depth of about 0.5–1 m from the ground surface to minimize the risk of low strength of stabilized soil and the water pollution in many cases, as shown in Figure 3.32. The original soil of the top of the stabilized soil column/element is heavily disturbed by the mixing blades to decrease its strength. The unstabilized soil at the shallow depth is usually excavated to the top of stabilized soil column/element before the construction of superstructure to prevent any adverse influence to the stability and settlement of the superstructure. As the excavated soil contains binder, it should be disposed appropriately according to the specification at the site.

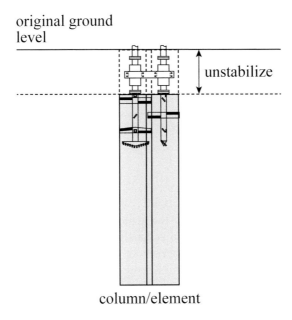

Figure 3.32 Production of stabilized soil column/element at shallow depth.

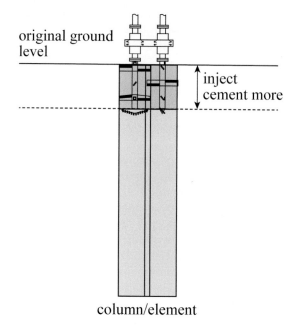

Figure 3.33 Production of stabilized soil column/element to ground surface.

In the case where stabilized soil columns/elements are produced to the ground surface by any reasons, additional amount of binder, as a compensation of the binder flowing out, should be injected in the ground at shallow depth to achieve the design strength there. In this case, excavation at shallow depth is not necessary for constructing superstructure (Figure 3.33).

It is desirable to construct a sand mat of about 0.5–1 m in thickness on the ground surface before the production to minimize the outflow of binder slurry by the enhanced overburden pressure by the sand mat. The stabilized soil column/element can be produced to the top of ground surface without additional binder injection at the shallow depth. Superstructure is constructed on the sand layer without excavation (Figure 3.34).

3.4.3.2 Influence of ground heaving

It should be noted that the original ground is heaved due to the stabilization. As a result of injecting binder into a ground, the ground around the stabilized soil columns/elements is displaced horizontally and verticality and the ground surface is heaved to some extent and accumulated by the series of production of stabilized soil columns/elements, as already shown in Figures 2.16 and 2.17. The ground level is already heaved up by the construction of previous stabilized soil column/element, "column/element A", before constructing a stabilized soil column, "column/element B", and will be heaved up by the production of the "column/element B", as shown in Figure 3.35. The stabilized soil column/element is usually produced to the depth of 0.5–1 m from the

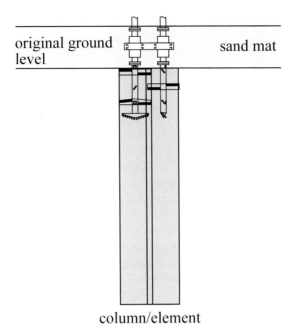

Figure 3.34 Production of stabilized soil column/element at shallow depth with sand mat.

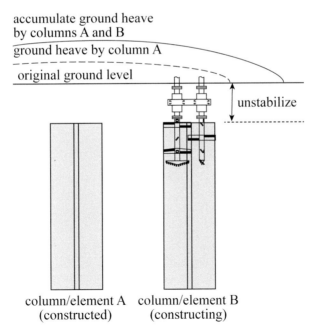

Figure 3.35 Production of stabilized soil columns/elements at shallow depth with ground heaving.

original ground surface irrespective of the amount of ground heaving, as shown in Figure 3.32. As the original soil on the stabilized soil column/element and the heaved soil are heavily disturbed, they are usually excavated to the top of stabilized soil column/element before the construction of superstructure to prevent any adverse influence to the stability and settlement of the superstructure, as shown in Figure 3.35. As the excavated soil contains binder, it should be disposed appropriately according to the specification at the site.

In the case where stabilized soil columns are produced to the heaved ground surface by any reasons, additional amount of binder, as a compensation of the binder flowing out, should be injected in the ground at shallow depth and heaved soil to achieve the design strength there. As the ground is heaved during the producing the column/element, it is quite difficult to produce the column/element to the heaved ground surface. In addition, the heaved ground level becomes higher and higher by the following production of columns/elements. The top level of the column/element becomes higher, and the top level of the previously produced column/element, "element A", is no longer the same as the ground after constructing the "element B" even if the "element A" was produced to the heaved ground surface. The ground level on the "element B" will be heaved up by the following columns/elements' production, which suggests that neither stabilized soil column/element reaches to the final heaved ground surface and heavily disturbed original soil exists on the columns/elements. The effect of the heaved soil on the columns/elements on the stability and settlement of superstructure should be evaluated when superstructure is constructed without excavation of heaved soil.

In the case of constructing a sand mat on the ground, a similar phenomenon can be seen to Figure 3.36. The height of ground level becomes higher and higher by the following production of columns/elements, as shown in Figure 3.37. As the result, neither stabilized soil column/element reaches to the final heaved ground surface and

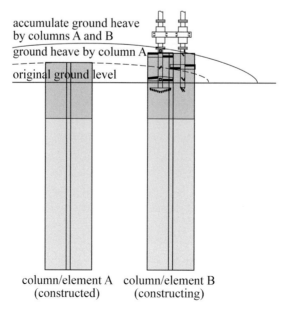

Figure 3.36 Production of stabilized soil columns/elements in heaved ground.

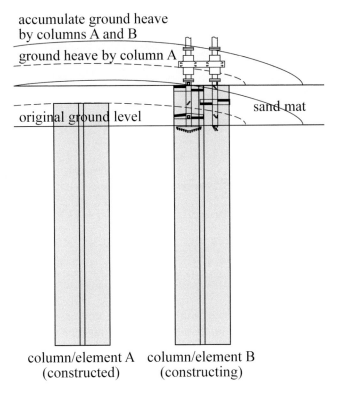

Figure 3.37 Production of stabilized soil columns/elements in heaved ground with sand mat.

heavily disturbed original soil exists on the columns/elements. The effect of the heaved soil on the columns/elements on the stability and settlement of superstructure should be evaluated when superstructure is constructed without excavation of heaved soil.

3.4.4 Bottom treatment

At the bottom of stabilized soil column/element, careful bottom mixing process by repeating penetration and withdrawal while injecting binder is usually conducted to attain the sufficient level of mixing. The quality of bottom end of column/element is critical especially in the case of the fixed type improved ground. As shown in Figure 3.38, there are two injection nozzles on the mixing shaft in many deep mixing equipment, where binder is injected from the bottom injection nozzle in the case of the penetration injection and from the top injection nozzle in the case of the withdrawal injection. The binder is injected from the bottom nozzle for the bottom treatment. The bottom injection nozzle is not installed at the tip of the mixing shaft but installed at slightly higher position than the tip in many deep mixing equipment. In addition, the binder slurry is not injected downward but horizontal direction as the nozzle is installed of the side surface of the shaft. Therefore, it is potentially possible

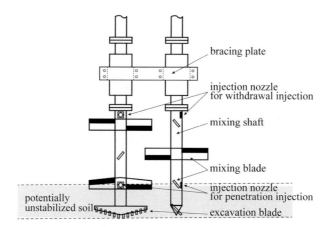

Figure 3.38 Production of stabilized soil column/element at bottom (bottom treatment).

that the soil between the tip of mixing shaft and the injection nozzle is disturbed without binder (see Figure 3.38).

Appropriate production processing is necessary not to leave the soft clay layer beneath the stabilized soil column/element for achieving tight connection of the stabilized columns/elements to the stiff layer. Several measures are taken, such as injecting more binder with/without time interval, and up and down movements with injecting binder, etc. However, as they are not always effective for any ground condition, it is necessary to find an appropriate solution by trial and error in the field-trial test.

3.4.5 Overlap columns/elements

As described in Chapter 2, it is specified to complete the overlap of stabilized soil columns within 24 hours from the previous produced column/element in order to achieve the tight connection of the columns/elements (Cement Deep Mixing Method Association, 1999). However, it is sometimes difficult to complete the overlap within the specified time at site under high temperature condition, where the strength of soil and binder mixture is increased rapidly due to the enhanced binder hydration. Many research studies were carried out to develop additives for suppressing the binder hydration and strength gain of several days after the mixing (Nishida and Sugita, 1997, Inada and Matsui, 1999, Kiyota et al., 2003, Okabe et al., 2011).

Figure 3.39 shows an example of effect of additive on the strength gain (Nishida and Sugita, 1997). In the test, three types of soil, soil A (sandy soil, sand particle content of 87.8%, silt particle content of 8.0% and clay particle content of 4.2%), soil B (silty soil, w_l of 32.2% and w_p of 21.1%, sand particle content of 35.9%, silt particle content of 43.9% and clay particle content of 20.2%) and soil C (clay soil, w_l of 79.8% and w_p of 43.4%, sand particle content of 0.6%, silt particle content of 38.6% and clay particle content of 60.8%) were stabilized with ordinary Portland cement of the cement factor of 200 kg/m^3 and w/c of 100% together with oxycarboxylic acid additive. In the soil A, the strength of the stabilized soil with the additive is smaller than that without

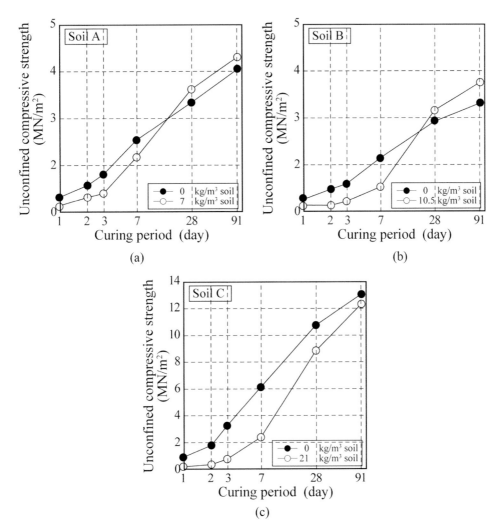

Figure 3.39 Effect of additive on strength increase of stabilized soil (Nishida and Sugita, 1997). (a) Soil A. (b) Soil B. (c) Soil C.

additive up to 7 days and larger after. A similar phenomenon can be seen in the soil B and soil C, while the effect of the additive is more dominant in the silt soil (soil B) and clay soil (soil C).

A similar effect was found for blast furnace slag cement type B and special cement and for various amounts of cement. Figure 3.40a shows the effect of amount of additive for the soil B (Nishida and Sugita, 1997), where the amount of additive was changed from 0 to 21 kg per 1 m^3 of soil. The 'days late' on the vertical axis is defined as days subtracted 1 day from the day when the strength gain of stabilized soil with the additive becomes the same as 1 day strength of stabilized soil without the additive. The figure shows that the 'days late' is less than 3 days as long as the amount of the additive

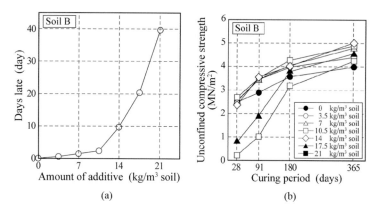

Figure 3.40 Effect of additive on strength increase of stabilized soil (Nishida and Sugita, 1997). (a) Days late and amount of additive. (b) Strength increase as curing period.

is less than 10.5 kg/m³ but is increased rapidly with the increase in the amount of the additive. Figure 3.40b shows the effect of additive on the long-term strength for the soil B. The long-term strength of the stabilized soil is almost the same order as that of stabilized soil without the additive when the amount of the additive is less than about 14 kg/m³. In the case of amount of the additive of 17.5 and 21 kg/m3, the strength is smaller than that of stabilized soil without the additive till 91 days but is almost the same as that without the additives after 180 days.

3.5 VERIFICATION TECHNIQUES IN QUALITY ASSURANCE

3.5.1 Core boring

3.5.1.1 Procedure

After the improvement, the quality of the field-stabilized soil columns/elements should be verified in advance of the construction of superstructure in order to confirm the design quality, such as continuity, uniformity, strength, permeability or dimension of stabilized soil column/element. A variety of verification test procedures to examine the engineering characteristics of stabilized soil have been proposed, as already shown in Table 2.4 (Hosoya et al., 1996; Halkola, 1999; Larsson, 2005). However, actual practices rely upon traditional verification techniques such as the unconfined compression test on boring core samples and/or the column penetration test in Nordic countries. The unconfined compression test on boring core sample is admitted as the best technique. Most of the other techniques seem to be used only for the research purpose or for settling the non-compliance. This may be due to their unfamiliarity both to owner/designer and contractor. Another reason may be the lack of direct correlation between their measured data and the design parameters, such as strength and stiffness.

3.5.1.2 Frequency of boring core sampling and specimen

As already mentioned in Section 2.3.8 of Chapter 2, the boring core samples should be obtained from each representative site and soil layer. In the case of large project, the improved area can be spatially classified into several zones depending on the ground condition, improvement condition (mixing condition) and superstructure condition. The boring core samples are taken from each inspection zone and inspection layer, as shown in Figure 2.15 so that the statistical evaluation can be conducted. The frequency of core sampling is determined depending on the size of project and the construction condition with various criteria such as the number, length and volume of stabilized soil column/element.

In marine works in Japan, the boring core is sampled every $10,000\,m^3$ for small-size project and every $50,000\,m^3$ for large-scale project exceeding $100,000\,m^3$ in stabilized soil volume. In the case of on land works in Japan, three core borings are generally conducted in the case where the total number of columns/elements is less than 500. When the total number exceeds 500, one additional core boring is conducted for every further 250 columns/elements, which corresponds about 0.5% of total production elements when the number of production column/element is around 1,000–2,000. The boring core is taken from the top to bottom through the inspection stabilized soil column/element irrespective of the marine and on land works. Three core barrels of 1 m length are selected from the boring core samples at three different depths and three specimens are taken from each core barrel and subjected to an unconfined compression test.

The Federal Highway Administration (FHWA) also specifies the frequency of boring core as 3% of total production elements on typical Deep Mixing project, which is about six times larger than the Japanese specification (Federal Highway Administration, 2013). The FHWA specifies the frequency of unconfined compression test of at least 5 for each full depth core, which is less than the Japanese specification.

3.5.1.3 Coring boring technique

There are many types of core boring techniques for various soil types (The Japanese Geotechnical Society, 2013). As the quality of boring core sample is influenced by the type of sampler. Figure 3.41 shows an example of boring core sample of rock by the ordinal soil sampler and triple tube sampler. The figure clearly shows the quality of boring core sample is influenced by the type of boring technique, in which it is hard to find specimen for laboratory test from the heavily broken core sample. Therefore, suitable core boring technique should be selected for the stabilized soil. As the strength of stabilized soil by the deep mixing method is the order of 1–$10\,MN/m^2$ in many cases, the double tube sampler and triple tube sampler are desired, while the double tube sampler has been frequently applied in Japan (Cement Deep Mixing Method Association, 1999). The double tube sampling with the boring core diameter of 86 mm is usually used in on land works, in which boring core sample of 65–68 mm in diameter can be sampled. In marine works, the double tube sampling with the boring core diameter of 116 mm is used, in which the sample of 95–98 mm in diameter can be sampled.

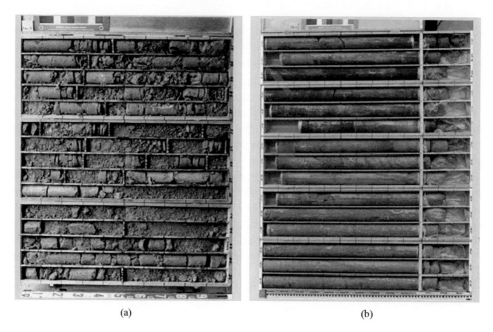

Figure 3.41 Boring core samples of rock (https://www.kiso.co.jp/services/development/boring-sampling.html). (a) By ordinal boring sampler. (b) By triple tube sampler.

3.5.1.4 Size of boring core

As mentioned in Section 3.3, the size of specimen influences the magnitude and *COV* of unconfined compressive strength in laboratory mix test. Here, the effect of specimen size on the strength of field-stabilized soil is introduced.

Figures 3.42 shows an example of the effect of core size on the strength of stabilized soil and *RQD* of boring core sample obtained in the field test (Enami et al., 1986b). In the field test, Funabashi loam (w_p of 81.3% and w_l of 146.3%) was stabilized with special cement of the cement factor of $300\,kg/m^3$ by a small-size deep mixing equipment. The specimen for the unconfined compression test was prepared by trimming the boring core sample that has the same diameter of the boring core sample and the height of the two times of the specimen diameter.

Figure 3.42a shows the effect of core size on the strength of stabilized soil. Though there is a lot of scatter in the data of the 50 mm sample, the unconfined compressive strength of the 50 mm in diameter shows about 45% larger strength than 100 mm sample, while the strength of 75 mm in diameter is almost the same strength of those of 100 mm size sample. This shows that the strength is decreased with the increase in the specimen size, which is a similar phenomenon to the laboratory mix specimen, as already shown in Figures 3.6–3.8. Figure 3.42b shows the effect of core size on *RQD* of boring core sample. The *RQD* of 50 mm size sample shows a lot of scatter, which might be influenced by the particle size of original soil. It may cause a lot of scatter in the

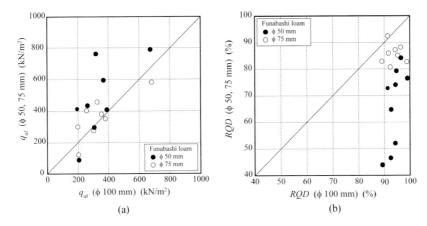

Figure 3.42 Effect of size of core boring on strength and *RQD* of stabilized soil (Enami et al., 1986b). (a) Unconfined compressive strength. (b) *RQD* of stabilized soil.

unconfined compressive strength. Based on the test results, they recommended the size of core boring should be equal or larger than 75 mm in diameter.

The diameter of specimen for the unconfined compression test is usually the same as that of the core boring sample in Japan, 65–68 mm for the on land works and 95–98 mm for the marine works. The height of specimen is selected two times of the specimen diameter.

3.5.1.5 Macroscopic evaluation of strength of field-stabilized soil

As mentioned in Section 3.3, the average strength and variation of unconfined compressive strength of stabilized soil are influenced by the size of specimen in laboratory mixed specimen that can be assumed to have relatively uniformity. It is found that the unconfined compressive strength and *COV* are decreased with the increase in the specimen size in general. In the case of field-stabilized soil, however, it is not homogenate and has a lot of scatter in the properties even if the stabilized soil column/element is produced with the best care and quality control. There are a lot of research efforts to investigate the cause behind the inhomogeneity of field-stabilized soil. Moreover, there are many research efforts by the field test and numerical calculations to evaluate the strength of full-scale stabilized soil column/element from the macroscopic view point (Omine et al., 2005).

Figure 3.43 shows an example of the relationship between the unconfined compressive strength on the full-scale stabilized soil column and the average strength on small-size specimen cored from the column for various soil types (Enami et al., 1986b; Enami and Hibino, 1991). Figure 3.43a (Enami et al., 1986b) shows the relationship on the Konosu clay (w_l of 149.2% and w_p f 50.1%, and organic matter content of 9.34%) and the Funabashi loam (w_l of 146.3% and w_p of 81.3%, and organic matter content of 17.66%) stabilized with special cement with various cement factors and *w/c* ratios. The figure shows the relationship between the unconfined compressive strength of

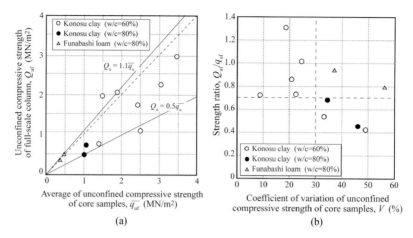

Figure 3.43 Relationship between unconfined compressive strengths of full-scale column and of small-size core specimen (Enami et al., 1986b). (a) Full-scale column and small core specimen. (b) Strength ratio and COV.

full-scale column, Q_u, and the average strength on small-size core specimen, q_{uf}. It is found that the Q_u has a linear relation with the q_{uf} irrespective of the type of soil and the Q_{uf}/q_{uf} ranges 0.5–1.1. Figure 3.43b shows the relationship between the strength ratio, Q_u/q_{uf}, and the coefficient of variation of the strength small-size core specimen, COV. It is also found that the Q_u/q_{uf} is decreased with the increase in the COV.

Enami and Hibino (1991) carried out a field test on various types of soil and mixing conditions, where the soil was stabilized by the small-size deep mixing equipment with the mixing blade of 400, 500 and 600 mm in diameter, where the cement factor and w/c ratio of cement slurry were changed from 250 to 350 kg/m³ and 60%–120% respectively, and the deep mixing equipment was also changed to investigate their effect on the column strength. Figure 3.44 shows the relationship between the unconfined compressive strength of full-scale column, Q_u, and the average strength on small-size core specimen, q_{uf}. It is found that the Q_u has a linear relation with the q_{uf} irrespective of the type of soil and the Q_{uf}/q_{uf} about 0.75 irrespective of the soil type and deep mixing equipment.

Figure 3.45 shows a similar relationship between the unconfined compressive strength on the full-scale stabilized soil column and the average strength of small-scale specimen cored from the column for various soil conditions and deep mixing equipment (The Building Center of Japan and the Center for Better Living, 2018). The Building Center of Japan and the Center for Better Living conducted a series of compression tests on cement stabilized soils excavated at 26 sites in order to investigate the size effect on the strength. The original soils are classified into five types, as shown in Figure 3.46 (The Building Center of Japan, 1997). In the fields, the stabilized soil columns were produced by the wet method of deep mixing equipment, in which the cement factor was 210–220 kg/m³ and the water and cement ratio of cement slurry, w/c was either 60% or 100%. In the tests, the field-stabilized soil columns excavated from the fields and trimmed to make a specimen with the diameter of about 1.0–1.2 m

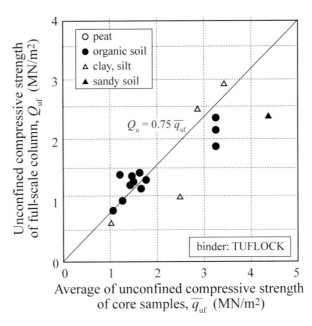

Figure 3.44 Relationship between unconfined compressive strength of full-scale column and average strength of small core specimen (Enami and Hibino, 1991).

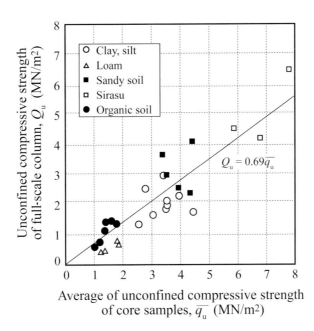

Figure 3.45 Relationship between unconfined compressive strength of full-scale column and average strength of small core specimen (The Building Center of Japan, 1997).

Figure 3.46 Soil sampler for wet grab sampling (Hashimoto et al., 2009). (a) Molds. (b) Stabilized soil in mold.

and height of about 1.5–2.4 m in height for the unconfined compression test. Small-size specimen with 67 mm in diameter and 130 mm in height were also sampled from the boring core of the field-stabilized soil column. The coefficient of variation of the strength on the small-size specimens ranges from 12.4% to 57.3%, and 38.0% in average. Figure 3.45 compares the unconfined compressive strength on the full-scale column, Q_u, and the average of unconfined compressive strength on the core specimen, q_{uf}. It is found that the Q_u has a linear relation with the q_{uf} irrespective of the type of soil and the Q_{uf}/q_{uf} about 0.69 irrespective of the soil type.

According to the accumulated data shown in Figures 3.43–3.45, the unconfined compressive strength of the full-scale column, Q_u, has a linear relation with the average strength on small-size core specimen, q_{uf}, and the Q_{uf}/q_{uf} is about 0.7–0.75 as far as the COV of q_{uf} is around 20%–40% and the strength ratio is decreased with the increase in the COV of q_{uf}.

Itou and Horiuchi (1980) investigate the effect of specimen size on the ratio of Young's modulus to the unconfined compressive strength, E_{50}/q_u, where the ratio, E_{50}/q_u, of the full-scale stabilized soil column is about 300–360 that is almost the same as that of the boring core sample.

3.5.2 Applicability of wet grab sampling

3.5.2.1 Type of wet grab sampling

The quality assurance (QA) is usually conducted by the unconfined compression test on boring core samples taken from the stabilized soil columns/elements after a prescribed curing period in Japan. One of the other approaches of the QA is the wet grab sampling, in which the stabilized soil mixture produced in-situ is sampled and molded soon after the production, and cured and tested in a similar manner as in laboratory mix test. Advantages of the wet grab sampling is that it allows deep mixing execution and sampling to be conducted in parallel, and sparing the effort to mobilize equipment on another day again.

There are two types of wet grab sampling: (a) sampling and filling stabilized soil mixture in a mold underground and (b) sampling stabilized soil mixture underground and filling it in a mold on ground. The latter case has been frequently applied for the preliminary QA of the cutter soil mixing method (see Figure 3.5).

In the former case, Figure 3.46 shows an example of mold with 70 mm in diameter and 140 mm in height and stabilized soil (Hashimoto et al. 2008, 2009). The molds installed on the rod are penetrated in a fresh stabilized soil column/element to the prescribed depth into the ground to take the soil and binder mixture there, as shown in Figure 3.46a. The mixture in the mold is lifted to trim its both sides and is cured in a laboratory for the strength test.

In the latter case, on the other hand, the soil sampler installed on the mixing shaft or mixing blade is penetrated into the fresh stabilized soil column/element to the prescribed depth (Figure 3.47a and b). The soil in the sampler are lift up to ground to fill in a mold on ground, in which the stabilized soil mixture is often mixed by hand before molding by hand to make a homogeneous sample. Figure 3.47c shows the other type of sampler that is installed on the bottom of rod to penetrate in a ground.

In both cases, the soil mixture filled in the mold is cured in a laboratory and subjected to strength test after the prescribed curing period. The soil conditions in the mold such as uniformity and density may be different from that in the underground especially in the latter case, and the curing condition such as temperature and

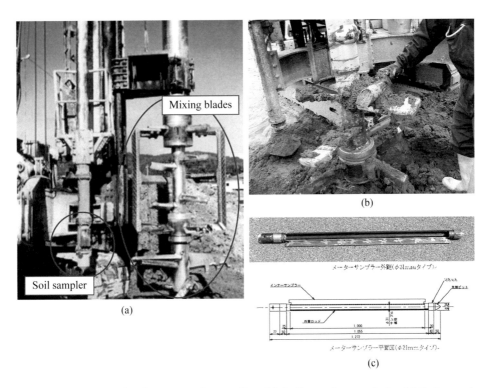

Figure 3.47 Soil sampler for wet grab sampling. (a) Soil sampler on shaft. (b) Soil sampler on mixing blade. (c) Soil sampler.

overburden pressure is also different from that in the underground. Therefore, it can be said that the strength on the wet grab sample is different from that on the boring core sample that is produced and cured underground.

3.5.2.2 Comparison of sampling type

Figure 3.48 shows an example of the effect of the sampling type on the strength of stabilized peat soils by the trencher mixing method (Power Blender Method Association, 2018), where the peat soft soil was stabilized by special cement with the cement factor of 70 kg/m^3 and w/c of 100% for Wakkanai-A soil and 76 kg/m^3 and w/c of 100% for Wakkanai-B soil (Hashimoto et al., 2008). The specimen for unconfined compression test were prepared by three techniques; "Core" corresponding boring core specimen after curing underground, "Mold" corresponding filling the mixture in a mold underground (Figure 3.46) and "Surface slurry" corresponding filling a mold with the mixture flowing out to the ground surface (Hashimoto et al., 2009).

In the case of the Wakkanai-A soil (Figure 3.48a), the "Core" shows a large scatter with the average strength of 308 kN/m^2. The "Mold" shows some amount of scatter in the measured data depend on the sampling depth, but ranging from about 300 to 400 kN/m^2. The "Surface slurry" shows a little larger strength than the "Core" and "Mold". In the case of the Wakkanai-B soil (Figure 3.48b), the "Core" shows a large scatter and very small average strength, which is the authors estimated due to quite low *RQD* of around 60%. The "Surface slurry" shows smaller than the "Mold," which is opposite phenomenon in the case of the Wakkanai-A soil. The unconfined compressive strength on the "Mold" and "Surface slurry" are different depending on the site; the strength on "Surface slurry" is a little larger than the "Mold" for the Wakkanai-A soil but quite smaller for the Wakkanai-B soil. By comparing them to the "Core," the "Core" is almost the same order as the "Mold" and "Surface slurry" for the Wakkanai-A soil but quite smaller for the Wakkanai-B soil.

Figure 3.49 shows the other example of the comparison of the two wet grab samplings, which is obtained by the above mentioned two methods (The Building Center of Japan and the Center for Better Living, 2018). The grounds were stabilized by small-size wet method deep mixing equipment with the cement factor of about

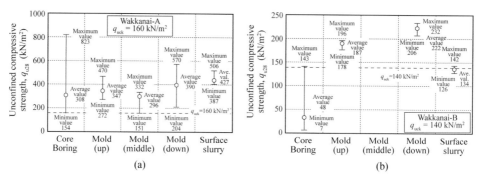

Figure 3.48 Comparison of molding type on q_{uf} (Hashimoto et al., 2009). (a) Wakkanai-A soil. (b) Wakkanai-B soil.

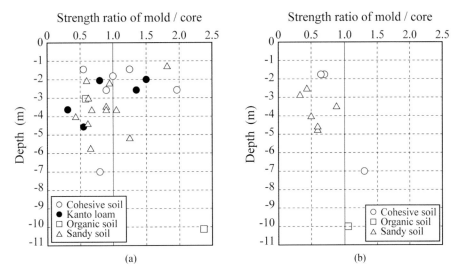

Figure 3.49 Strength ratio of molding soil strength to boring core strength (The Building Center of Japan and the Center for Better Living, 2018). (a) Molding on ground. (b) Molding in underground.

150 to 200 kg/m³ to achieve the design strength of the order of 500 kN/m² in unconfined compressive strength. For the sampling type (2), filling on ground, Figure 3.49a, the strength ratio, $q_{u,\,mold}/q_{u,\,core}$, is scattered ranging from about 0.3 to 2.0 depending on the soil type and the mixing condition. For the sampling type (1), filling underground, Figure 3.49b, the strength ratio is about 0.4–0.8 in the depth to −5 m and about 1.0–1.3 in the depth to −7 to −10 m irrespective of the soil type.

As far as the literature survey, it cannot be concluded at the present stage about the clear relationship with the two methods, filling soil and binder mixture underground and filling mixture on ground. It is necessary to accumulate further data to obtain a clear tendency of the sampling types.

3.5.2.3 Comparison of wet grab sample strength and boring core sample strength

1. Deep mixing method

    Figure 3.50 shows the accumulated data on the comparison of the strength of boring core sample, q_{uf} and the wet grab sample (fill soil and binder mixture on ground), q_{uw}, in which the wet and dry methods are plotted together (Kitazume and Nishimura, 2009; 2012; Ryan and Jasperse, 1989; Bertero et al., 2015; Leoni and Shcmutzler, 2015; Chen and Driller, 2012). Although there is a lot of scatter in the test data in Figure 3.50a depending on the site, soil type, equipment type and mixing condition, a linear relationship can be seen in each test data in general. Figure 3.50b shows the strength ratio, $q_{u,\,wet}/q_{u,\,core}$, along with the boring core strength, $q_{u,\,core}$. Though there is also a lot of scatter, it can be seen roughly that the strength ratio is decreased in general with the increase in the boring core strength.

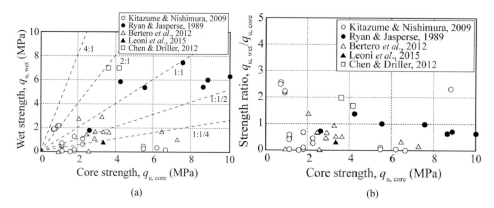

Figure 3.50 Comparison of strengths of boring core sample and wet grab sample for deep mixing method. (a) Boring core and wet grab strengths. (b) Strength ratio.

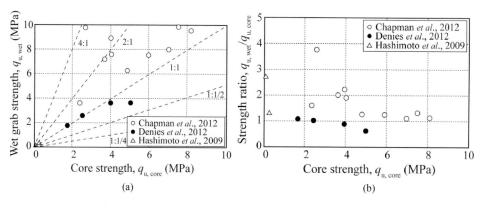

Figure 3.51 Comparison of strengths of boring core sample and wet grab sample for continuous mixing method. (a) Boring core and wet grab strengths. (b) Strength ratio.

2. Cutter soil mixing method

Figure 3.51 shows an example of the relationship between the strength on the boring core sample and the wet grab sample for the cutter soil mixing method (CSM). There is a lot of scatter in the data depending on the construction site, as shown in Figure 3.51a. Figure 3.51b shows the strength ratio, $q_{u,\,wet}/q_{u,\,core}$. There is a lot of scatter in the test data depending on the construction site, especially for the low strength range, lower than about 4 MN/m^2. However, it can be seen that the strength ratio is decreased in general with the increase in the boring core strength.

3.5.2.4 Applicability of wet grab sampling for QA

Applicability of the wet grab sampling techniques was explored as a potentially useful complementary method in quality assurance of the deep mixing method. The overall range of the strength ratio of wet grab sample strength to boring core strength, $q_{u,\,wet}/q_{u,\,core}$, was around 0.1–2.5, as shown in Figures 3.50b and 3.51b. As the range is still wide and scattered depending on the construction site, further discussion on the applicability of the wet grab sampling is required with accumulating more case studies. According to the test results, the wet grab sampling can be applicable to estimate the strength of boring core sample roughly, however, the correlation between the wet grab and boring core strengths should be obtained for each soil and project. In the case of the CSM method, the web grab sampling technique can be applicable for estimating the field strength precisely in the case where the core strength is larger than around 4 MN/m^2.

REFERENCES

Abe, T., Miyoshi, A., Maeda, T. and Fukuzumi, H. (1997) Evaluation for soil improvement of clay mixing consolidation method using dual-way mixing system. *Proceedings of the 32nd Annual Conference of the Japanese Geotechnical Society*, pp. 2353–2354 (in Japanese).

Åhnberg, H. and Holm, C. (2009) Influence of laboratory procedures on properties of stabilised soil specimen. *Proceedings of the International Symposium on Deep Mixing and Admixture Stabilization*, pp. 167–172.

Åhnberg, H. and Holm, G. (1984) Triaxial testing of naturally cemented carbonate soil. *Swedish Geotechnical Institute Report*. No. 30, pp. 93–146. (in Swedish).

Aoi, M., Komoto, T. and Ashida, S. (1996) Application of TRD method to waste treatment on the ground. *Proceedings of the 2nd International Congress on Environmental Geotechnics, IS-Osaka '96*, pp. 437–440 (in Japanese).

Babasaki, R., Terashi, M., Suzuki, T., Maekawa, A., Kawamura, M. and Fukazawa, E. (1996) Japanese Geotechnical Society Technical Committee Reports: Factors influencing the strength of improved soil. *Proceedings of the 2nd International Conference on Ground Improvement Geosystems*. Vol. 2, pp. 913–918.

Bertero, A., Morales, C., Leoni, F.M. and Filz, G. (2015) Three deep mixing projects Comparison between laboratory and field test results. *Proceedings of the Deep Mixing 2015*, San Francisco, June 2–5, pp. 699–714.

Cement Deep Mixing Method Association (1999) *Cement Deep Mixing Method (CDM), Design and Construction Manual*. Cement Deep Mixing Method Association, 192p. (in Japanese).

Chen, W.Y. and Driller, M. (2012) Results of a cement deep soil mixing test section at Perris dam. *Proceedings of the 4th International Conference on Grouting and Deep Mixing*.

Coastal Development Institute of Technology (2019) *Technical Manual of Deep mixing method for marine works, revised version*. Daikousha Publishers, 315p. (in Japanese).

Enami, A. and Hibino, S. (1991) Improvement of foundation ground by using a deep soil blender with free blades. *Journal of the Japanese Society of Soil Mechanics and Foundation Engineering, "Tsuchi to Kiso"*. Vol. 39, No. 10, pp. 37–42 (in Japanese).

Enami, A., Yoshino, M., Hibino, S., Takahashi, M. and Akiya, K. (1985a) In situ measurement of temperature in soil cement (improved soil) columns and influence of curing temperature on unconfined compressive strength of soil cement. *Proceedings of the 20th Annual Conference of the Japanese Society of Soil Mechanics and Foundation Engineering*, pp. 1737–1740 (in Japanese).

Enami, A., Yoshino, M., Hibino, S., Takahashi, M. and Akiya, K. (1985b) Basic properties of soil cement columns produced by Teno Column method (Deep mixing method) using free blades. *Proceedings of the 20th Annual Conference of the Japanese Society of Soil Mechanics and Foundation Engineering*, pp. 1755–1758 (in Japanese).

Federal Highway Administration (2013) *Federal Highway Administration Design Manual_Deep Mixing for Embankment and Foundation Support.*

Fujii, M., Kawamura, M., Tamura, M., Watanabe, K. and Mizoguchi, E. (2004) Quality evaluation method for soil-cement column by deep mixing method. *Journal of the Society of Materials Science, Japan.* Vol. 53, No. 1, pp. 9–12 (in Japanese).

Grisolia, M., Leder, E. and Marzano, I.P. (2013) Standardization of the molding procedures for stabilized soil specimens as used for QC/QA in deep mixing application. *Proceedings of the 18th International Conference on Soil Mechanics and Geotechnical Engineering.*

Halkola, H. (1999) Keynote lecture: Quality control for dry mix methods. *Proceedings of the International Conference on Dry Mix Methods for Deep Stabilization.* Stockholm, pp. 285–294.

Hashimoto, H., Nishimoto, S. and Hayashi, H. (2008) Variation in strength of improved ground by trencher mixing method and the quality control method. *Proceedings of the Annual Research Meeting, Hokkaidou Branch, The Japanese Geotechnical Society.* Vol. 48, pp. 219–224 (in Japanese).

Hashimoto, H., Nishimoto, S. and Hayashi, H. (2009) Investigation of improvement strength variation for the trencher mixing method. *Proceedings of the International Symposium on Deep Mixing and Admixture Stabilization.*

Hayashi, N., Ochiai, H., Yasuhuku, N. and Omine, K. (1996) Size effect on splitting test and unconfined compression test of cement-treated soils. *Proceedings of the 2nd National Symposium on Ground Improvement,* pp. 245–250 (in Japanese).

Hirabayashi, H., Taguchi, H., Tokunaga, S., Shinkawa, N., Fujita, T., Inagawa, H. and Yasuoka, N. (2009) Laboratory mixing tests on cement slurry preparation, specimen preparation and curing temperature. *Proceedings of the International Symposium on Deep Mixing and Admixture Stabilization,* pp. 141–144.

Horpibulsuk, S., Miura, N. and Nagaraj, T.S. (2003) Assessment of strength development in cement-admixed high water content clays with Abrams' law as a basis. *Geotechnique.* Vol. 53, No. 4, pp. 439–444.

Hosoya, Y., Nasu, T., Hibi, Y., Ogino, T., Kohata, Y. and Makihara, Y. (1996) Geotextile tubes and beneficial reuse of dredged soil_ applications near ports and harbours. *Proceedings of the 2nd International Conference on Ground Improvement Geosystems.* Vol. 2, pp. 919–924.

Isobe, K., Samaru, Y., Aoki, C., Sogo, K. and Murakami, T. (1996) Large scales deep soil mixing and quality control. *Proceedings of the 2nd International Conference on Ground Improvement Geosystems.* pp. 619–624.

Inada, K. and Matsui, K. (1999) Properties of soil cement mixture added superretarder. *Proceedings of the 34th Annual Conference of the Japanese Geotechnical Society,* pp. 853–854 (in Japanese).

Itou, M. and Horiuchi, S. (1980) Ground improvement effect by deep mixing method - Strength profile of stabilized soil column and full scale column strength. *Proceedings of the 15th Annual Conference of the Japanese Society of Soil Mechanics and Foundation Engineering,* pp. 1769–1772 (in Japanese).

Jeong, G.H., Shin, M.S., Han, G.T., Lee, J.H. and Kim, J.H. (2009) Studying for lab mixing test of task 2 in Korea. *Proceedings of the International Symposium on Deep Mixing and Admixture Stabilization,* pp. 145–149.

Kido, Y., Nishimoto, S., Hayashi, H. and Hashimoto, H. (2009) Effects of curing temperatures on the strength of cement-treated peat. *Proceedings of the International Symposium on Deep Mixing and Admixture Stabilization,* pp. 151–154.

Kitazume, M. and Nishimura, S. (2009) Influence of specimen preparation and curing conditions on unconfined compression behaviour of cement-treated clay. *Proceedings of the International Symposium on Deep Mixing and Admixture Stabilization,* pp. 155–160.

Kitazume, M. and Nishimura, S. (2012) Applicability of molding procedures in laboratory mix tests for quality control and assurance of the deep mixing method. *Proceedings of the 4th International Conference of Grouting and Deep Mixing*, pp. 427–436.

Kitazume, M. and Terashi, M. (2009) International collaborative study on QA/QC for deep mixing -proposal. *Proceedings of the International Symposium on Deep Mixing and Admixture Stabilization*, pp. 103–105.

Kitazume, M. and Terashi, M. (2013) *The Deep Mixing Method*. CRC Press, Taylor & Francis Group, Boca Raton, FL, 410p.

Kitazume, M., Grisolia, M., Leder, E., Marzano, I.P., Correia, A.A.S., Oliveira, P.J.V., Åhnberg, H. and Andersson, M. (2015) Applicability of molding procedures in laboratory mix tests for quality control and assurance of the deep mixing method. *Soils and Foundations*. Vol. 55, No. 4, pp. 761–777.

Kitazume, M., Nishimura, S., Terashi, M. and Ohishi, K. (2009a) International collaborative study task 1: Investigation into practice of laboratory mix tests as means of QC/QA for deep mixing method. *Proceedings of the International Symposium on Deep Mixing and Admixture Stabilization*, pp. 107–126.

Kitazume, M., Ohishi, K., Nishimura, S. and Terashi, M. (2009b) International collaborative study task 2 report Interpretation of comparative test program. *Proceedings of the International Symposium on Deep Mixing and Admixture Stabilization*, pp. 127–140.

Kiyota, M., Tutumi, T., Ohta, K. and Hirayama, Y. (2003) About the characteristic of the improvement soil with the slow-setting soil stabilizer. *Proceedings of the 38th Annual Conference of the Japanese Geotechnical Society*, pp. 799–800 (in Japanese).

Kusakabe, F., Maeda, T. and Kawanabe, O. (1996) Laboratory model test for the deep mixing pile method and the type of mixing blade. *Proceedings of the 31st Annual Conference of the Japanese Geotechnical Society*, pp. 135–136 (in Japanese).

Larsson, S. (2005) State of Practice Report - Execution, monitoring and quality control. *Proceedings of the International Conference on Deep Mixing – Best Practice and Recent Advances*, Stockholm. Vol. 2, pp. 732–785.

Leoni, F.M. and Shcmutzler, W.D. (2015) Wet grab samples vs cored samples as QC/QA methods at the Rio Puerto Nuevo project in San Juan PR. *Proceedings of the International Foundations Conference and Equipment Expo*, 2015, pp. 2420–2431.

Marzano, I.P., Al-Tabbaa, A. and Grisolia, M. (2009) Influence of sample preparation on the strength of cement-stabilised clays. *Proceedings of the International Symposium on Deep Mixing and Admixture Stabilization*, pp. 161–166.

Miyazaki, Y., Tang, Y.X., Ochiai, H., Yasufuku, N. and Omine, K. (2001) A correlation among unconfined compressive strength, cement content and water content for cement treated dredging. *Technology Reports of Kyushu University*, Vol. 74, No. 1, pp. 1–8 (in Japanese).

Mizuno, S., Sudou, F., Kawamoto, K. and Endou, S. (1986) Ground displacement due to ground improvement by deep mixing method and countermeasures. *Proceedings of the 3rd Annual Symposium of the Japan Society of Civil Engineers on Experiences in Construction*, pp. 5–12 (in Japanese).

Nakama, T., Saitou, S. and Babasaki, R. (2003) Early stage estimation for 28-day unconfined compressive strength of cement-improved soils using an accelerated curing test. *Proceedings of the 38th Annual Conference of the Japanese Geotechnical Society*, pp. 797–798 (in Japanese).

Nakamura, M. and Matsushita, M. (1982) Studies on "mixing efficiency" of chemicals for deep mixing method. *Proceedings of the 17th Annual Conference of the Japanese Society of Soil Mechanics and Foundation Engineering*, pp. 2597–2600 (in Japanese).

Nakamura, M., Matsuzawa, S. and Matsushita, M. (1982) Studies on "mixing efficiency" of stirring wings for deep mixing method. *Proceedings of the 17th Annual Conference of the Japanese Society of Soil Mechanics and Foundation Engineering*, pp. 2585–2588 (in Japanese).

Nishibayashi, K., Matsuo, T., Hosoya, Y. and Kohinata, T. (1985a) Studies of strength and deformation properties of soils stabilized with cement (Part 3). *Report of Obayashi Corporation Technical Research Institute.* No. 30, pp. 123–127 (in Japanese).

Nishibayashi, K., Matsuo, T., Hosoya, Y. and Umetsu, K. (1985b) Improvement of deep soft ground by cement mixing (Part 7). *Report of Obayashi Corporation Technical Research Institute.* No. 31, pp. 114–118 (in Japanese).

Nishida, M. and Sugita, S. (1997) Fundamental study of the cement-improved soil using retarder. *Proceedings of the 32nd Annual Conference of the Japanese Geotechnical Society,* pp. 2441–2442 (in Japanese).

Noto, S., Kuchida, N. and Terashi, M. (1983) Case histories of the deep mixing method. *Journal of the Japanese Society of Soil Mechanics and Foundation Engineering, "Tsuchi to Kiso".* Vol. 31, No. 6, 7, pp. 73–80 (in Japanese).

Okabe, M., Kawamura, K., Masuda, K. and Seki, T. (2011) Ingenious device of site management in ground improvement work by CI-CMC method. *Journal of the Japanese Society of Soil Mechanics and Foundation Engineering, "Tsuchi to Kiso".* pp. 16–19 (in Japanese).

Omine, K., Ochiai, H. and Yasufuku, N. (2005) Evaluation of scale effect on strength of cement-treated soils based on a probabilistic failure model. *Soils and Foundations.* Vol. 45, No. 3, pp. 125–134.

Omine, K., Ochiai, H. and Yoshida, N. (1998) Estimation of in-situ strength of cement-treated soils based on a two-phase mixture model. *Soils and Foundations.* Vol. 38, No. 4, pp. 17–29.

Omura, T., Murata, M. and Hirai, N. (1981) Site measurement of hydration-generated temperature in ground improved by deep mixing method and effect of curing temperature on improved soil. *Proceedings of the 36th Annual Conference of the Japan Society of Civil Engineers,* pp. 732–733 (in Japanese).

Power Blender Method Association (2018) *Technical manual of power blender method.* Power Blender Association. (in Japanese).

Public Works Research Center (2004) *Technical manual on deep mixing method for on land works.* Public Works Research Center, 334p. (in Japanese).

Ryan, C.R. and Jasperse, B.H. (1989) Deep soil mixing at the Jackson lake dam. *ASCE Geotechnical and Construction Divisions Special Conference,* pp. 1–14.

Saito, J., Nishibayashi, K. and Matsuo, T. (1981a) Improvement of deep soft ground by cement mixing (Part 2). *Report of Obayashi Corporation Technical Research Institute.* No. 22, pp. 110–114 (in Japanese).

Saito, J., Nishibayashi, K., Matsuo, T. and Hosoya, Y. (1981b) Improvement of deep soft ground by cement mixing (Part 3). *Report of Obayashi Corporation Technical Research Institute.* No. 23, pp. 87–91 (in Japanese).

Sakuma, S. (2013) Construction machine for soil cement mixing wall and deep soil mixing, Development, expansion and improvement of CSM method machine. *Japan Construction Machinery and Construction Association, Journal of Construction Machinery,* pp. 75–79 (in Japanese).

Satou, T. (2009) Development and application of mid depth mixing method - Power Blender method. *Kisoko,* Sougou Doboku Kenkyusho. Co., Ltd. Vol. 35, No. 5, pp. 57–59 (in Japanese).

Suzuki, M., Taguchi, T., Fujimoto, T., Kawahara, Y., Yamamoto, T. and Okabayashi, S. (2005) Effect of loading condition during curing period on unconfined compressive strength of cement stabilized soil. *Journal of Geotechnical Engineering, Japan Society of Civil Engineers.* Vol. 792, pp. 211–216 (in Japanese).

Tang, Y.X., Miyazaki, Y. and Tsuchida, T. (2000) Advanced reuses of dredging by cement treatment in practical engineering. *Proceedings of the International Conference of the Coastal Geotechnical Engineering in Practice.* Vol. 1, pp. 725–731.

Terashi, M. and Kitazume, M. (2009) Keynote Lecture: Current Practice and Future Perspective of QA/QC for Deep-Mixed Ground. *Proceedings of the International Symposium on Deep Mixing and Admixture Stabilization*, pp. 61–99.

The Building Center of Japan (1997) *Design and Quality Control Guideline of Improved Ground for Building*. The Building Center of Japan, 473p. (in Japanese).

The Building Center of Japan and the Center for Better Living (2018) *Design and Quality Control Guideline of Improved Ground for Building, 2018*. The Building Center of Japan and the Center for Better Living, 708p. (in Japanese).

The Japanese Geotechnical Society (2009) *Practice for making and curing stabilized soil specimens without compaction, JGS T 0821–2009*. The Japanese Geotechnical Society, 1156p. (in Japanese).

The Japanese Geotechnical Society (2013) *Geotechnical and Environmental Investigation Methods*. The Japanese Geotechnical Society. Vol. 2, 1259p. (in Japanese).

Tsuchida, T. and Tang, Y.X. (2012) A consideration on estimation of strength of cement-treated marine clays. *Japanese Geotechnical Journal*. Vol. 7, No. 3, pp. 435–447 (in Japanese).

Yamamoto, T. and Miyake, M. (1982) Influence of specimen size on unconfined compressive strength of cement treated soil. *Journal of the Society of Materials Science, Japan*. Vol. 31, No. 341, pp. 66–69 (in Japanese).

Yamamoto, T., Suzuki, M., Okabayashi, S., Fujino, H., Taguchi, T. and Fujimoto, T. (2002) Unconfined compressive strength of cement-stabilized soil cured under an overburden pressure. *Journal of Geotechnical Engineering, Japan Society of Civil Engineers*. Vol. 701, pp. 387–399 (in Japanese).

Yanagihara, M., Horiuchi, S. and Kawaguchi, M. (2000) Long-term stability of coal-fly-ash slurry man-made island. *Proceedings of the Coastal Geotechnical Engineering in Practice*, pp. 763–769.

Yoshida, T., Kubo, H. and Sumida, K. (1977) Investigation on treatment of sludge (Part 3) Application of pF water to relationship between strength of cement treated soil to water cement ratio. *Proceedings of the 12th Annual Conference of the Japanese Society of Soil Mechanics and Foundation Engineering*, pp. 1309–1312 (in Japanese).

Yoshizu, T. (2014) Development of excavating agitator in deep mixing soil method. *Architectural Institute of Japan, Summaries of Technical Papers of Annual Meeting*. Vol. 20, No. 44, pp. 25–28 (in Japanese).

Chapter 4

Problems and countermeasures associated with problematic soils

4.1 PROBLEMATIC SOIL FOR STABILIZATION

The field stabilized soil column/elements have relatively large strength variability even if the production is done with the established deep mixing equipment and procedure and with the best care since the strength is influenced by many factors. One of the reasons behind the strength variability is the type of soil. Some cohesive soils are so sticky that the entrained mixing phenomenon is happened, a condition in which disturbed soil adheres to and rotates with the mixing blade without efficient mixing, as shown in Figure 2.7. Table 4.1 summarizes the characteristics of problematic soil which is hard to be mixed well (Kawamura et al., 2001).

A soil with high fines content shows large short-term strength of soil and binder mixture and low fluidity, which cause low degree of mixing soil and binder (Aoyama et al., 2002). A high-activity soil, such as montmorillonite clay, absorbs water to cause insufficient and inhomogeneous mixing (Nozu et al., 2012, 2015). The fluidity of soil and binder mixture is also influenced by the type of soil mineral. The soil having high cation exchange capacity and high electrical conductivity has high adsorptive property of calcium ion to reduce the fluidity of soil and binder mixture (Aoyama et al., 2002). Several countermeasures against such a problematic soil are introduced.

Table 4.1 Characteristics of soil hard to be mixed (Kawamura et al., 2001).

Property	Criteria
Gravel content	Less than 5%
Fines content	Less than 80%
Consistency index	Larger than 0.2
Activity	High
Plasticity index	Larger than 40
Liquid limit	Larger than 80%
Sensitivity	Low

4.2 COUNTERMEASURES FOR PROBLEMATIC SOILS

4.2.1 Water injection

During the penetration, the mixing blades are rotating to disaggregate and disturb the soil to reduce the strength of ground so as to make the mixing blades penetrate by their self-weight. When encountering a stiff, cohesive and sticky soil layer, water is

DOI: 10.1201/9781003223054-4

sometimes injected from the bottom injection nozzle of the mixing shaft for softening the layer in some parts of the world. A similar procedure is also adopted in the jet grouting method, where water is injected in the first penetration and withdrawal stage, and binder slurry is injected in the second production stage. The water injection may be effective to soften original soil and increase its fluidity, but its effect is limited for some types of soil. However, the water injection always causes the decrease in the strength of stabilized soil and increase the ground horizontal movement and ground heaving and during the production of columns/elements. The influence of water injection on the strength of stabilized soil and countermeasure for compensating the water injection are discussed on the water to cement (W/C) ratio concept in the following sections.

4.2.2 Use of new type special cement

Aoyama et al. (2002) investigated the effect of the physico-chemical property of soil on the strength of stabilized soil and produced two special cements, named Cement-A and Cement-B, to stabilize cohesive soils that suppress the short-term strength increase but increase the long-term strength: Cement-A prevents the electrostatic repulsion by negative charging in order to keep the fluidity, and Cement-B prevents the electrostatic repulsion and polyvalent metal ion. The Cement-A and Cement-B are mixed with the two intermediate soils excavated at Tokorozawa, soil T, and excavated at Tokyo respectively, soil V. The soil T (fines content less than 5 μm, Fc of 49% and fines content less than 75 μm, Fs of 66%) contains a large amount of clay minerals and has large cation exchange capacity, and the soil V (Fc of 53% and Fs of 99%) contains large amount of polyvalent metal ions.

Figure 4.1a shows the shear strength of the fresh stabilized soil mixture measured by a hand vane apparatus. In the figure, the w/c and Vm/V are the water to cement ratio of binder slurry, and the volume ratio of the binder, Vm, to soil, V, respectively. The strength of soil and ordinary Portland cement mixture is increased soon after mixing, which corresponds to the decrease in the fluidity of soil and binder mixture. The soil and the Cement-A mixture shows almost no strength gain until 60 minutes

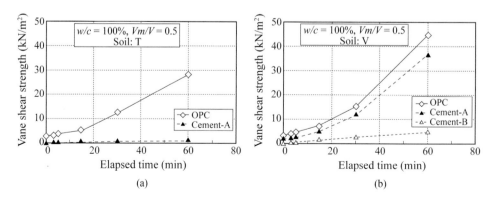

Figure 4.1 Strength increase in soil and cement mixture (Aoyama et al., 2002). (a) Fresh stabilized soil, T. (b) Fresh stabilized soil, V.

after mixing, which indicates that the fluidity of soil and binder mixture is kept high. A similar phenomenon can be seen in Figure 4.1b, where the strength of soil and ordinary Portland cement mixture is increased, but the soil and Cement-B mixture shows a slight increase in strength. It should be noted that the soil V stabilized with the Cement-A also shows large strength gain, while a slight strength increases was observed in the case of the soil T. The figures indicate that the strength increase in the fresh stabilized soil mixture can be suppressed by adding chemical substances suitable for soil type, which will be effective for increasing the mixing degree by maintaining high fluidity of soil and cement mixture.

4.2.3 Use of dispersant

Hirano et al. (2017) investigated the effect of two dispersants on the fluidity of soil and cement mixture in the laboratory mix test, where two types of soil, (a) artificial cohesive soil composed of Kaolin clay (70%) and quartz sand No. 5 (30%) and (b) artificial sandy soil composed of Kaolin clay (40%) and quartz sand No. 5 (60%) were stabilized with a special cement (Japan Cement Association, 2012) with the binder content of aw of 10% and w/c of cement slurry of 60% together with two type of dispersants, gluconic acid dispersant and polycarboxylic acid dispersant. The shear strength of the fresh soil and cement mixture for the cohesive soil measured by a hand vane apparatus is shown in Figure 4.2. In the case of the gluconic acid dispersant (Figure 4.2a), the shear strength of the fresh soil cement mixture remained quite small value irrespective of its amount and is almost constant within 60 minutes, while the shear strength of the mixture without dispersant is increased rapidly with the elapsed time. In the case of the polycarboxylic acid dispersant (Figure 4.2b), the strength of the mixture is smaller than that of mixture without the dispersant and is increased with the elapsed time. According to these figures, the gluconic acid dispersant and polycarboxylic acid dispersant function to keep the fluidity of the soil and binder mixture high soon after mixing.

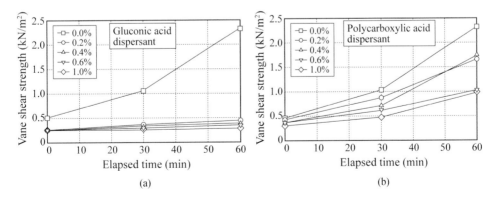

Figure 4.2 Effect of dispersant on strength increase in cohesive soil and binder mixture (Hirano et al., 2017). (a) Gluconic acid dispersant. (b) Polycarboxylic acid dispersant.

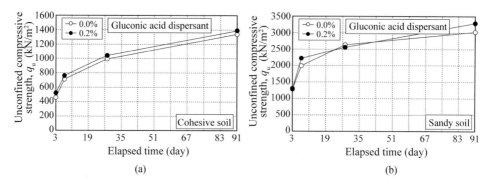

Figure 4.3 Effect of dispersant on strength increase of stabilized soil (Hirano et al., 2017). (a) Cohesive soil. (b) Sandy soil.

Figure 4.3 shows the unconfined compressive strength of the stabilized soil up to 91 days (Hirano et al., 2017). In the case of the cohesive soil, the unconfined compressive strength is increased with the elapsed time irrespective of the amount of dispersant. Similar behavior can be seen in the case of the sandy soil. It can be concluded that the strength of stabilized soil is not influenced by the amount of dispersant irrespective of the soil type.

According to the test results, the gluconic acid dispersant and polycarboxylic acid dispersant function to keep the fluidity of the soil and binder mixture high soon after the mixing and show negligible influence on the strength of stabilized soil.

Mizutani and Makiuchi (2003) carried out field mixing tests to investigate the effect of the surfactant additive on the mixing degree and strength of stabilized soil, in which the test conditions are summarized in Table 4.2. After production of the stabilized soil columns by the small size deep mixing equipment, the boring core samples were taken for the unconfined compression test. The unconfined compression test results are also summarized in Table 4.2 together. The average strength of the stabilized soil with the additive at 28 days is slightly smaller than that without the additive irrespective of the soil type, but the COV of the stabilized soil with the additive is smaller

Table 4.2 Test condition and unconfined compression test result.

		Stabilized column							UCS test	
Name	Soil type	Diameter (m)	Length (m)	Binder factor (kg/m³)	w/c (%)	Additive, Cw (%)	No. of columns	No. of Sample	Ave. UCS, (kN/m²)	COV, (%)
A-1	Sandy	0.6	3.0	280	60	0.0	2	5	5395.1	31.0
A-2	silt	0.6	3.0	280	60	1.0	3	5	5520.1	23.4
B-1	Kanto	0.6	3.0	300	65	0.0	2	5	2894.6	31.7
B-2	loam	0.6	3.0	300	65	1.0	3	5	2610.6	20.4
C-1	Sand	0.6	3.0	250	65	0.0	2	5	1093.4	50.5
C-2		0.6	3.0	250	65	1.0	3	5	1024.1	26.0
D-1	Clay	0.6	3.0	300	65	0.0	2	5	2874.9	54.1
D-2		0.6	3.0	300	65	2.0	3	5	2477.0	37.9

After Mizutani and Makiuchi (2003).

than that without the additives, which clearly indicates that the surfactant additive is effective to increase the mixing degree to achieve the homogeneity of stabilized soil column.

Similar phenomenon was reported on the soil in Vietnam and the USA, where clay containing montmorillonite absorbs water to cause insufficient mixing with soil and binder. Nozu et al. (2012, 2015) carried out the laboratory mix tests to investigate the effect of the two chemical admixtures on the fluidity of the fresh soil and binder mixture, as shown in Table 4.3. In the test, marine clay at the south Vietnam were stabilized with cement with the cement factor, Fc of $160\,kg/m^3$ and w/c of 100% by changing the type and amount of admixture, Admixture A (for retarding water reducing) and B (for increase liquidity for cement mix soil) and 0%, 1% and 2%, respectively. Figure 4.4a shows the table flow test results measured soon after the mixing, where the table flow (fluidity) of a fresh stabilized soil mixture is increased with the increase in the amount of admixture. The effect of admixture depends on its type, in which Admixture A shows larger effect than Admixture B. It indicates that the high fluidity of a fresh soil and binder mixture mitigate the entrained mixing phenomenon to increase the mixing degree, which in turn provides the large long-term strength and high uniformity of the strength of stabilized soil. Figure 4.4b shows the 28 days' strength of stabilized soil with the admixture, the strength is increased with the increase in the amount of admixture irrespective of the type of admixture. The test clearly shows the high applicability of the admixture to prevent the entrained mixing phenomenon and in turn to increase the mixing degree and the strength.

Table 4.3 Admixture (Nozu et al., 2015).

	Admixture A	Admixture B
Producer/Product name	BASF, Pozzolith	KAO, Mighty
Purpose	Retarding, water reducing	Increase liquidity
Place of product	Almost all countries	Japan and Thailand

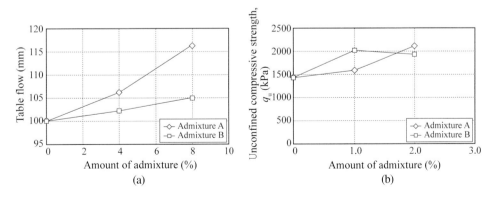

Figure. 4.4 Effect of admixture (Nozu et al., 2015). (a) Table flow of soil and binder mixture. (b) 28 days' unconfined compressive strength.

4.2.4 Injecting atomized cement slurry

New type of deep mixing equipment was developed, where cement slurry is accompanied with compressed air at injection nozzle and atomized cement slurry is injected to mitigate the entrained mixing phenomenon, as shown in Figure 4.5 (Murakami et al., 2015; Murakami, 2017). The injected air babbles can function as a ball bearing to increase the fluidity of soil and binder mixture. Figure 4.6 shows the strength distribution of stabilized soil within the element obtained in field tests, where sandy silt soil was stabilized with ordinary Portland cement to compare the atomized cement slurry injection technique (CI-CMC technique) and ordinary mixing technique. The 28 days' strength of stabilized soil element by the ordinary mixing technique shows comparatively small strength of the order of 1,000–1,700 kN/m^2, while that by the CI-CMC technique shows large strength of the order of 3,100–3,600 kN/m^2. The figure also shows the strength variation of stabilized soil element by the CI-CMC technique is smaller than that by the ordinary mixing technique. As the results, the atomized cement slurry injection can function to keep fluidity of the fresh stabilized soil mixture high and increase the strength and decrease the strength variation.

Atomized cement slurry can function to disturb the original soil structure and increase the fluidity of soil and cement mixture, which can contribute to the decrease in the penetration resistance and the required mixing energy and in turn the deep mixing equipment can penetrate in a hard soil layer. Figure 4.7 shows the measured auger torques to compare the CI-CMC technique and the ordinary mixing technique. In the case of the ordinary mixing technique, the penetration of the mixing tool is suspended by the hard layer at the depth of about −8m, while the CI-CMC technique can penetrate through the hard layer. According to the case history, the CI-CMC technique can penetrate about 2.5m into the mudstone layer having an SPT N-value of about 25–50 (q_u of 4–5 MN/m^2).

Figure 4.5 Deep mixing equipment with injecting atomized cement slurry (Murakami, 2017).

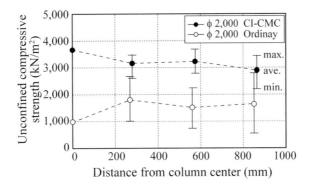

Figure 4.6 Strength distribution of stabilized soil within element (Murakami, 2017).

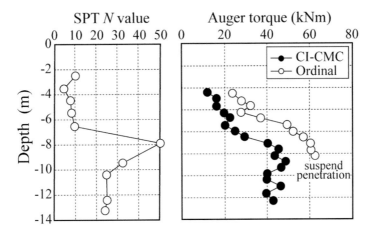

Figure 4.7 Effect of atomized cement slurry injection on penetration in hard layer (Murakami, 2017).

4.2.5 Summary

The approaches mentioned above require the installation of facilities and software on the deep mixing equipment and control system. In addition, it is necessary to study their effects and determine a suitable mixing condition in the laboratory mix test.

REFERENCES

Aoyama, K., Miyamori, T., Wakiyama, T. and Kikuchi, D. (2002) The influence of physical property on improved soil character. *Journal of the Japan Society of Civil Engineers.* Vol. 721/VI-57, pp. 207–219 (in Japanese).

Hirano, S., Mizutani, Y., Nakamura, H., Shimomura, S., Sasada, H. and Adachi, T. (2017) Laboratory mix tests of cement stabilized soil with dispersant. *Architectural Institute of Japan, Summaries of technical papers of the Kantou Branch* , pp. 1–4 (in Japanese).

Japan Cement Association (2012) *Soil improvement manual using cement stabilizer* (4th edition). Japan Cement Association, 442p. (in Japanese).

Kawamura, M., Hibino, S., Tamura, M., Fijii, M. and Watanabe, K. (2001) Perspective of performance-based deep-mixing method. *Journal of the Japanese Society of Soil Mechanics and Foundation Engineering, "Tsuchi to Kiso"*. Vol. 49, No. 5, pp. 1–3 (in Japanese).

Mizutani, Y. and Makiuchi, K. (2003) Effects of surfactant additive on strength stability of soil cement pile. *Proceedings of the 48th National Symposium on Geotechnical Engineering*, pp. 45–52 (in Japanese).

Murakami, S. (2017) Development of ejector type deep mixing method and its quality control system. Kisoko, Sougou Doboku Kenkyusho. Co., Ltd, pp. 44–47 (in Japanese).

Murakami, S., Ideno, T. and Asatsuma, R. (2015) Quality of stabilized column by the ejector type deep mixing method. *Japanese Geotechnical Society, Symposium on Ground, Disaster prevention and Environment*, pp. 1–6 (in Japanese).

Nozu, M., Anh, N.T., Shinkawa, N. and Matsushita, K. (2012) Remedy of deep soil mixing quality for montmorillonite clay deposited in the Mekong and Mississippi deltas. *ISSMGE-TC211*, Brussels. Vol. 2, pp. 443–449.

Nozu, M., Sakakibara, M. and Anh, N.T. (2015) Securing of in-situ cement mixing quality for the expansive soil with the Montmorillonite inclusion. *Proceedings of the Deep Mixing 2015*, San Francisco, June 2–5, pp. 845–852.

Chapter 5

Water to binder ratio concept in QC

5.1 INTRODUCTION

As already described, the quality of stabilized soil depends upon a number of factors including the type and condition of original soil, the type and amount of binder and the execution process. The quality control and quality assurance practice focuses upon the quality of stabilized soil, which comprises the laboratory mix test, field-trial test, monitoring and controlling construction parameters and verification. Construction control parameters during production include the continuous controlling and monitoring of penetration and withdrawal speeds of mixing tool, rotation speed of mixing blade, quantity of binder and water to binder ratio (for the wet method). The strength of stabilized soil is in principle influenced by the combination of water, soil and binder. The water to binder ratio is one of the key factors to evaluate the strength of stabilized soil. It is necessary to consider the water injection in the quality control. The importance of *W/C* ratio concept in the QC/QA is discussed in this chapter.

5.2 DEFINITION OF *W/C* RATIO

5.2.1 Definition of *W/C*

In the many admixture stabilization projects, the strength of field stabilized soil should be predicted and confirmed at various stages of planning, testing, design, and implementation. There are many proposed formulas to predict the laboratory strength and field strength of stabilized soil, which incorporate the various factors for the stabilization effect exemplified as Equation (3.1) (Tsuchida and Tang, 2012; Horpibulsuk et al., 2003; Tang et al., 2000; Yanagihara et al., 2000; Miyazaki et al., 2001; Yoshida et al., 1977). Among them, several key parameters can be found in the equations: weights of water, binder and soil. Here, one of the key parameters, the *W/C* ratio is introduced, which is defined as the total weight of water contained in soil and binder slurry to the weight of the binder (Equation 5.1). The ratio is also expressed by the water content of soil, binder content and water to binder ratio.

$$W/C = (W_{ws} + W_{wc})/W_c \\ = (w/aw + w/c) \tag{5.1}$$

DOI: 10.1201/9781003223054-5

where
- aw: binder content
- w: water content of soil
- w/c: water to binder ratio of binder slurry
- W/C: total water to binder ratio
- W_c: dry weight of binder (kg)
- W_{wc}: weight of water contained in binder slurry (kg)
- W_{ws}: weight of water contained in soil (kg)

5.2.2 Relationship between W/C ratio and stabilized soil strength

Figure 5.1 shows an example of the relationship between the W/C ratio and laboratory stabilized soil strength at 28 days' curing, q_{u28}, for various soils (Yoneda, 2011). The furnace slag cement type B was used as a binder in the laboratory mix tests, where the cement factor and the w/c ratio of cement slurry were changed to cover a wide range of the W/C ratio. In the figure, test case of the w/c ratio of 100% is shown, where the original soils are classified into three groups depending on its fines content, Fc: 0%–30%, 30%–70% and 70%–100%. It can be seen that the strength is rapidly decreased with the increase in the W/C ratio irrespective of the fines content. It is also found that the strength becomes larger when the fines content becomes large. Similar relationship has been found in various soils stabilized with various binders by the deep mixing method (Federal Highway Administration, 2013) and by the pneumatic flow mixing method (Kitazume and Satoh, 2003). According to the test results in Figure 5.1, the strength of stabilized soil can be formulated as Equation (5.2):

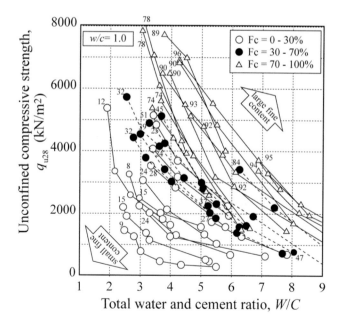

Figure 5.1 Relationship between W/C ratio and unconfined compressive strength of stabilized soil (Yoneda, 2011).

$$q_u = a \times (W/C)^b \tag{5.2}$$

where

a: parameter (kN/m^2)
b: parameter
q_u: unconfined compressive strength (kN/m^2)
W/C: total water to binder ratio

It is interesting to note that the parameter a in Equation (5.2) is larger for the soil with large fines content than that with small fines content. And the parameter b is a negative value and its absolute value becomes small with the decrease in the fines content. They indicate that the stabilized soil strength is large on soils with large fines content but is decreased with the increase in the W/C ratio. The W/C ratio influences the stabilized soil strength with large fines content.

5.3 PREDICTION OF FIELD STRENGTH BY PRODUCTION LOG DATA

5.3.1 Production log data

Figure 5.2 shows an example of deep mixing log data of production of an element in a marine work. The ground at the construction site consists of four layers, whose properties are summarized in Table 5.1 together with the sand mat, which was spread on the original ground to mitigate the water pollution during the production of element. The clay-1 and clay-2 layers are soft layer with the natural water content of around 65%–80%, while the Alluvial-clay and Alluvial-sand layers are relatively hard layer with low natural water content. The laboratory mix tests were carried out on the soils, and the

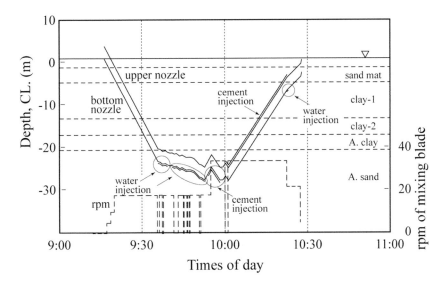

Figure 5.2 Example of deep mixing production log data.

Table 5.1 Ground condition

Depth		Soil name	Soil property			Estimation of strength, $a * (W/C)^b$	
Upper (m)	Lower (m)		Liquid limit, w_l (%)	Plastic limit, w_p (%)	Water content, w_n (%)	a (MN/ m^2)	b
−2.40	−4.40	Sand mat	0.0	0.0	20.0	37.8	−1.62
−4.40	−13.23	Clay-1	65.0	31.0	79.5	37.8	−1.62
−13.23	−16.85	Clay-2	65.0	31.0	65.0	37.8	−1.62
−16.85	−20.35	Alluvial-clay	58.0	28.0	54.0	16.7	−0.98
−20.35	−50.00	Alluvial-sand			53.0	16.7	−0.98

relationships between the *W/C* ratio by changing the cement dosage and the laboratory strength, q_{ul}, are examined to obtain parameters a and b, as shown in Table 5.1.

The mixing tool is equipped with two binder injection nozzles at the bottom and top mixing blades for the penetration injection and withdrawal injection, respectively, as exemplified schematically in Figure 5.3. In Figure 5.2, the positions of the bottom and top mixing blade and the rotation speed of the mixing blade are plotted along the time of day. In the production, as shown in Figure 5.2, the mixing blades were penetrated into the clay-1, clay-2 and Alluvial-clay layers at a constant speed of 1 m/min., while the rotation speed of mixing blade was kept almost constant of 18 rpm. When the bottom end of mixing tool reached to the Alluvial-sand layer at the depth of −23.2 m, the penetration speed was reduced to 0.13 m/min due to the hard Alluvial-sand layer where the required torque for rotating the mixing blade was jumped up. The mixing tool was continued to penetrate into the Alluvial-sand layer to the depth of −27.5 m with the low penetration speed, while the mixing blade was rotated at the constant speed with several intentional stops. During the penetration, water was injected about 1.1 m³ at the depth of −23.2 to −23.8 m and about 5.09 m³ at the depth of −24.7 to −27.5 m from the bottom injection nozzle for expecting softening the Alluvial-sand layer for ease of penetration. After reaching the design depth of −27.5 m, the mixing blades was gone up and down for the bottom treatment, while cement slurry of the *w/c* of 80% was injected from the bottom injection nozzle (penetration injection). However, as the delivery tube from the water pump to the bottom injection nozzle had been filled with water, as shown in Figure 5.3, the water in the delivery tube was injected first and then the cement slurry was injected. The water injection is not identified in the production log data. The volume of injected water and the required time for injection depend on the diameter and length of the delivery tube from the cement slurry pump to the injection nozzle and the flow rate of cement slurry pump, which can be roughly estimated about 0.5–1 m³ and 0.7–1.4 min., respectively. After the bottom treatment, the cement slurry was continuously injected from the top nozzle (withdrawal injection), while the rotation speed of the mixing blade was increased to 34 rpm. After one operation, the delivery tube was filled with the cement slurry. As the delivery tube from the cement slurry and water pumps to the mixing tool is shared for cement slurry and water, the water was injected from the bottom nozzle soon after the end of withdrawal injection in order to flush cement delivery tube for the next production. The flushing water was injected at −7.1 to −6.5 m in the clay-1 layer in this site, which caused the *W/C* ratio increase and the strength decrease there.

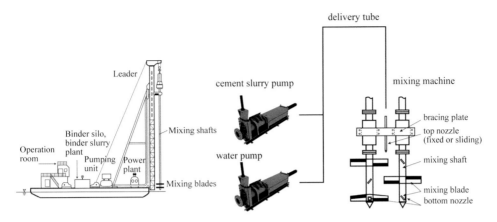

Figure 5.3 Schematic view of delivery tube from pump to mixing tool.

5.3.2 Analysis of production log data

The production log data shown in Figure 5.2 is analyzed to evaluate the mixing condition every one meter depth and the analyzed result is shown in Figures 5.4–5.8. Figure 5.4 shows the volume of injected cement slurry calculated based on the flow rate. The figure shows that it is almost constant of about $0.8\,m^3/m$ (about 17.6% of the soil volume to be stabilized) at the depth of -5 to $-20.35\,m$ but varied at the depth of -20.35 to $-27.5\,m$ in the Alluvial-sand layer for the bottom treatment. The volume of injected water was about 0.25–$1.8\,m^3/m$ at the depth of around -23 to $-24\,m$ and around 0.5–$1.0\,m^3$ at about the depth of $-26.6\,m$ in the Alluvial-sand layer and about $0.45\,m^3$ at about the depth of $-7\,m$ in the clay-1 layer for flushing the binder delivery tube. In the figure, the water remained in the tube before the cement injection is not considered.

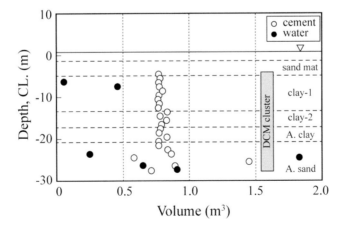

Figure 5.4 Distributions of volume of cement slurry and water along depth.

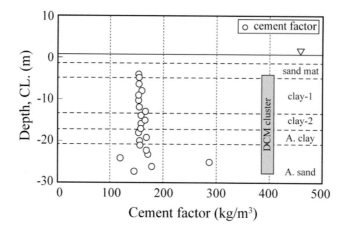

Figure 5.5 Distribution of binder factor along depth.

Figure 5.5 shows the distribution of the cement factor along the depth, which is defined as the weight of cement to the unit volume of original soil to be stabilized. The cement factor is almost constant of about 150 kg/m³ as the design criteria for the clay-1 and clay-2 and Alluvial-clay layers. In the Alluvial-sand layer, the cement factor is scattered ranging from 118 to 289 kg/m³, because the cement slurry and water were injected for the bottom treatment and flushing the delivery tube.

The *W/C* ratio is calculated by Equation (5.1) and plotted along the depth in Figure 5.6. In the clay-1 layer, the *W/C* ratio is almost constant of about 4.8 except at the depth of around −7 m, where the water of 0.48 m³ was injected for flushing the delivery tube after the withdrawal injection. In the clay-2 and Alluvial-clay layers, the *W/C* ratio is smaller than that of the clay-1 layer and even the amount of cement is almost the same as shown in Figure 5.5, which is due to the low natural water content of the layers rather

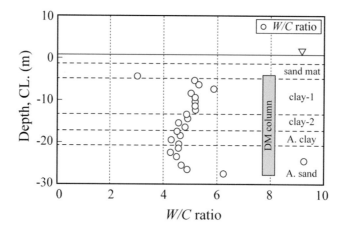

Figure 5.6 Distribution of W/C ratio along depth.

than that of the clay-1 layer, as shown in Table 5.1. The *W/C* ratio in the Alluvial-sand layer is scattered from 4.3 to 9.3, which is due to the scatter in the amount of injected cement slurry and water.

The blade rotation number, *BRN*, is calculated by Equation (2.3) and shown in Figure 5.7. The *BRN* is very large of about 600 as the number of mixing blades is large, and almost constant through the depth except the Alluvial-sand layer. In the Alluvial-sand layer, quite large *BRN* can be found ranging around 1230–2700 due to the bottom treatment.

As explained in the previous section, the unconfined compressive strength has a close relationship with the *W/C* ratio. The parameters *a* and *b* of Equation (5.2) had been obtained in the laboratory mix tests for each soil layer, as shown earlier in Table 5.1. The estimated unconfined compressive strength of the laboratory soil, q_{ul}, is shown along the depth in Figure 5.8. The predicted strength at the depth of about −7 m is about 2.1 MN/m², which is about 20% smaller than the others, which is due to the water injection there. In the Alluvial-clay layer, the predicted strength is almost constant of around 3.9 MN/m². In the Alluvial-sand layer, the predicted strength is decreased with the depth. The estimated strength at the depth of around −26.0 m is quite small of about 1.9 MN/m².

As shown in Figures 5.4–5.8, the laboratory strength, q_{ul}, can be estimated by the production log data during the production. And the field strength, q_{uf}, can also be estimated if incorporating the strength ratio, q_{uf}/q_{ul}.

The boring core sample was taken from the stabilized soil element, and the unconfined compression tests were carried out to measure the field strength, q_{uf}. These measured field strengths are plotted in Figure 5.8 by solid circles. The measured strengths in the clay-1 and clay-2 are quite smaller than the estimated strength. The ratio of the measured to estimated strength, $q_{uf}/q_{ul,\ est}$, is roughly obtained as about 0.63. It should be noted that the measured field strength is scattered rather than the estimated strength, which might be due to the variation of the local soil condition, mixing condition (entrained mixing phenomenon), and the boring core and unconfined compression test apparatus and techniques.

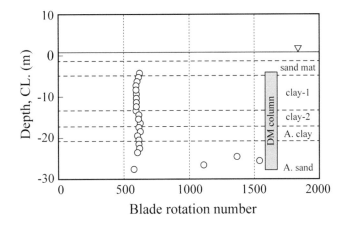

Figure 5.7 Distribution of blade rotation number, *BRN* along depth.

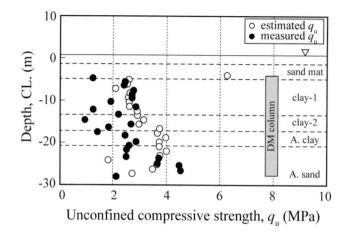

Figure 5.8 Distribution of predicted laboratory strength and measured field strength along depth.

According to the analysis of the production log data, it can be concluded that the construction procedure could be well organized and controlled, where the cement content and the blade rotation number assured the design criteria. However, the water injection at several depths and the variation of the natural water content along the depth might have caused the scatter in the estimated strength. In addition of the scatter in the mixing condition, boring core machine and technique and testing technique might cause the scatter in the strength.

5.3.3 Countermeasure for water injection

As described above, the *W/C* ratio has a close relationship with the unconfined compressive strength and can be applied in QC/QA during the production of element to assure the design strength. One of the possible applications can be introduced here by referring to the production log data, as shown in Figure 5.2. As discussed in the previous section, some amount of water was injected at the depth of about −23.2 to 23.8 m and −24.7 to −27.5 m for softening the hard layer and at the depth of about −27.7 to −25.9 m for replacing the water by the cement slurry. If, in the withdrawal stage, cement slurry was injected at the constant flow rate irrespective of the water injection, the *W/C* ratio at the layer subjected to water injection would become larger than that of the other layer resulting in smaller unconfined compressive strength. Additional cement slurry should be injected to the layers for compensating the water injection. The amount of cement slurry to be injected can be calculated for assuring the same *W/C* ratio and the same strength, while Equation (5.1) can be modified as Equation (5.3) for incorporating the water injection. In this particular case, the volume of additional cement slurry to be injected can be calculated, as shown in Table 5.2:

Table 5.2 Additional cement slurry to be injected for compensating water injection

Depth	Volume of water injected (m³)	Volume of additional cement slurry to be injected (m³)	Ratio of additional to original cement volume	Average
−23.2 to −23.8 m	1.1	0.33	69%	57%
−24.7 to −27.5 m	5.09	1.51	68%	
−27.7 to −26.7 m	0.5 to 1.0*	0.15 to 0.3	19%–38%	
−7.1 to −6.5 m	0.5–1.0*	0.13–0.26	27%–54%	

*Estimated.

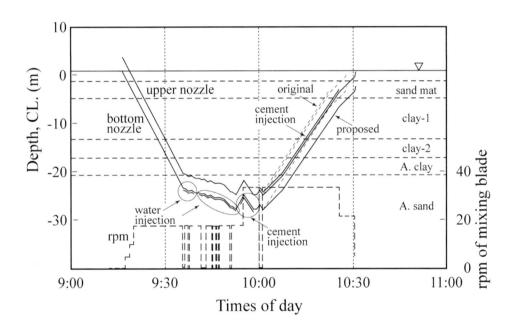

Figure 5.9 Proposed deep mixing construction procedure for compensating water injection.

$$W/C = (W_{ws} + W_{wc} + W_{ww})/W_c \\ = (w/aw + w/c) + W_{ww}/W_c \quad (5.3)$$

where
  W_{ww}: weight of water injected (kg)

It is rather easy in practice to change the withdrawal speed for injecting more cement rather than increase the flow rate of cement slurry in the target soil layer temporarily. Figure 5.9 shows a proposed construction diagram, where the withdrawal speed at the depth of −27.7 to −23.2 m in the Alluvial-sand layer is reduced to 0.57 m/min for injecting additional cement slurry to assure the design strength.

5.4 SUMMARY

Construction control parameters during the production of columns/elements in the deep mixing include the continuous controlling and monitoring of penetration and withdrawal speeds of mixing tool, rotation speed of mixing blade, quantity of binder and water/binder ratio (for the wet method). It is well known that the field stabilized soil column/element has relatively large strength variability even if the execution is done with the established deep mixing equipment and with the best care. One of the reasons behind the strength variability is the type of soil. Some cohesive soils are so sticky that the soil and binder mixture adheres to and rotates with the mixing blade without efficient mixing. For such soils, several alternative approaches are introduced in the manuscript: using new type special cement, injecting chemical additives and injecting air.

When encountering stiff, cohesive and sticky soil, water is sometimes injected from the bottom injection nozzle for increasing fluidity of original soil. The water injection may be effective to increase the fluidity of original soil, but its effect is limited for some types of soil. However, the water injection always causes the decrease in the strength of stabilized soil. The influence of water injection on the strength and countermeasure for compensating the water injection is discussed based on the water to cement (W/C) ratio concept.

The W/C ratio has a close relationship with the unconfined compressive strength and can be applied in QC/QA during construction to assure the design strength. The field stabilized soil strength can be estimated by the production log data during the production. Once predicting the strength during the production, several countermeasures can be adopted and carried out during the production to assure the design criteria. In the manuscript, one of the countermeasures is introduced for compensating the water injection, where the volume of additional cement slurry to be injected is calculated to assure the design strength. This concept can be applied to the case where the water content of soil layer is different locally.

In recent years, Information Communication Technology (ICT) has been applied in the construction industry to increase productivity and reliability of construction. The ICT construction machines are featured with the advanced technology such as machine guidance system and execution control system to assist the operator, and the site management system to process productivity and work progress data.

As the ICT and AI technologies are developed very rapidly, ICT deep mixing equipment will be working at site to produce the stabilized soil columns/elements in near future. In the machine, the type and properties of the ground profile are estimated by the measured construction control parameters, such as hanging load, penetration speed of mixing tool, driving torque of mixing blade and so on in the penetration stage, and the appropriate mixing condition, such as amount of binder and blade rotation number and so on are obtained by the help of the W/C concept and the accumulated data base together with the AI technology to achieve the design strength along the depth. The stabilized soil column/element is produced automatically by ICT deep mixing equipment according to the obtained mixing condition.

REFERENCES

Federal Highway Administration (2013) *Federal Highway Administration Design Manual_Deep Mixing for Embankment and Foundation Support.*
Horpibulsuk, S., Miura, N. and Nagaraj, T.S. (2003) Assessment of strength development in cement-admixed high water content clays with Abrams' law as a basis. *Geotechnique.* Vol. 53, No. 4, pp. 439–444.
Kitazume, M. and Satoh, T. (2003) Development of pneumatic flow mixing method and its application to Central Japan International Airport construction. *Journal of Ground Improvement.* Vol. 7, No. 3, pp. 139–148 (in Japanese).
Miyazaki, Y., Tang, Y.X., Ochiai, H., Yasufuku, N. and Omine, K. (2001) A correlation among unconfined compressive strength, cement content and water content for cement treated dredging. *Technology Reports of Kyushu University.* Vol. 74, No. 1, pp. 1–8 (in Japanese).
Tang, Y.X., Miyazaki, Y. and Tsuchida, T. (2000) Advanced reuses of dredging by cement treatment in practical engineering. *Proc. of the International Conference of the Coastal Geotechnical Engineering in Practice.* Vol. 1, pp. 725–731.
Tsuchida, T. and Tang, Y.X. (2012) A consideration on estimation of strength of cement-treated marine clays. *Japanese Geotechnical Journal.* Vol. 7, No. 3, pp. 435–447 (in Japanese).
Yanagihara, M., Horiuchi, S. and Kawaguchi, M. (2000) Long-term stability of coal-fly-ash slurry man-made island. *Proceedings of the Coastal Geotechnical Engineering in Practice*, pp. 763–769.
Yoneda, K. (2011) Investigation of strength property of stabilized soil. *Geotechnical Forum*, (in Japanese).
Yoshida, T., Kubo, H. and Sumida, K. (1977) Investigation on treatment of sludge (Part 3) Application of pF water to relationship between strength of cement treated soil to water cement ratio. *Proceedings of the 12th Annual Conference of the Japanese Society of Soil Mechanics and Foundation Engineering*, pp. 1309–1312 (in Japanese).

Index

additive 21, 86, 108
admixture 2, 109
axial strain at failure 48

bearing capacity 2, 11
bending strength 16
binder 2
    content 25, 107
    factor 24, 54, 118
    slurry 19, 25, 58, 113
blade
    excavation blade 28, 76
    free blade 7, 26, 75
    mixing blade 6, 26, 33, 73, 75, 76
blade rotation number 34, 78, 119
bottom treatment 18, 29, 35, 85, 117

calcium 2, 105
cement 2
    blast furnace slag cement 23, 39, 87
    content 61
    factor 39
    hydration 2
    ordinary Portland cement 60, 86
    special cement 31, 65, 68, 73, 77, 79, 87, 90, 95, 106
    stabilization 2
coefficient of variation (COV) 21, 55, 65
cold joint 32
column/element 6, 7, 9, 20, 44, 55, 88
    non-compliant column/element 19, 49
combination of mechanical and high pressure injection mixing 4
curing
    condition 15, 23, 57
    period 68, 71, 95
    temperature 67, 68, 71, 73

deep mixing 4, 54
    CDM method 4, 32, 80
    CDM-LODIC method 39
    CI-CMC method 110
    construction equipment 39, 73
    CSM method 59, 98
    DJM method 4, 34, 80
    dry method 2, 5, 34, 54
    equipment 7, 26, 73, 122
    JACSMAN method 5
    Power Blender method 59
    TRD method 59
    wet method 2, 5, 34, 54
densification 2
density
    of binder slurry 41
    of stabilized soil 58, 64
design
    geotechnical design 4, 15, 17, 22
    process design 15, 18, 21, 22, 54, 57
dewatered stabilized soil 3

E_{50}/q_u 62, 94
entrained mixing phenomenon 6, 26, 76, 105
ex-situ mixing 3

field trial test 15, 18, 27, 54
fixed type improvement 16, 28, 35, 85
floating type improvement 16, 28
fluidity
    of binder 23
    of soil and binder mixture 58, 64, 105

geotechnical design 4, 15, 17, 22
grain size distribution 20
ground heaving 25, 3, 41, 81, 106
grouting 2

Index

high pressure injection mixing 4, 5

improved ground 10, 15, 17, 44, 53
improvement
    area ratio 11, 38
    block type improvement 10, 11, 32, 35, 68
    fixed type improvement 16, 28, 35, 85
    floating type improvement 16, 28
    grid type improvement 10, 11, 35
    group column type improvement 10, 11, 31
    pattern 11, 18
    tangent column type improvement 11
    type 10
    wall type improvement 10, 11, 39
injection method 28, 34
    penetration injection method 28, 30, 33, 34
    withdrawal injection method 28, 30, 33, 34, 76
in-situ mixing 3, 4
installation pattern 9, 10, 16
ion exchange 2, 105, 106

laboratory mix test 16, 23, 25, 54, 59, 119
lime 2
    lime column method 4
liquefaction 3, 4, 11
    prevention 11
    resistance 4

marine work 7, 35, 45, 54, 89
maturity 71
mechanical mixing 3, 4
mixing
    blade 6, 24, 33, 73, 75, 105
    combination of mechanical and high pressure injection mixing 4
    condition 18, 23, 57, 117
    degree 20, 23, 34, 61, 73, 80, 108
    equipment 6, 9, 26, 34, 39, 49, 54, 73
    ex-situ mixing 3
    high pressure injection mixing 4, 5
    in-situ mixing 3, 4
    mechanical mixing 3, 4
    shaft 6, 27, 33, 58, 73, 81
    tool 9, 28, 37, 73, 77, 116

non-compliant column/element 19, 49

on land work 5, 33, 45, 54, 89
organic matter content 20, 68, 91
overlap 10, 30, 34, 86

permeability 18, 47
pH 20
pozzolanic reaction 2
probability 21, 49
process control 15, 18
process design 15, 18, 21, 22, 54, 57
production log data 19, 35, 40, 49, 115, 117
production process 54, 86

quality assurance 5, 11, 15, 19, 88, 95, 98
quality control 4, 11, 15, 28, 36

rock quality designation (RQD) 46, 90, 97

soil and binder mixture 23, 58, 86, 105, 110
stability 10, 11, 32, 49, 81
    external stability 10
    internal stability 10
strength
    bending strength 16
    coefficient of variation (COV) 21, 55, 65
    design standard strength, q_{uck} 16, 20, 49
    field strength, q_{uf} 18, 20, 48, 53, 55, 115
    laboratory strength, q_{ul} 23, 25, 53, 54, 113
    long term strength 88, 106
    residual strength 48
    size effect 60, 92
    standard deviation 21, 48
    strength decrease 25, 81
    strength increase 65, 88
    strength ratio, q_{uf}/q_{ul} 22, 54, 58, 65, 119
    tensile strength 16, 60
    unconfined compressive strength 19, 20, 45, 54, 91, 115
    unconfined compressive strength of full scale column, Q_{uf} 92
surface treatment 3

table flow 109
tensile strength 16, 60
termination depth 19
thermal stabilization 2

verification technique 88
verticality 19, 35, 40, 82

water content 25, 48, 58, 113
water injection 25, 105, 116, 120, 122
water to binder ratio 19, 25, 28, 32, 37, 58, 113